D1333046

BETWEEN LIGHT AND STORM

Also by Esther Woolfson

Corvus: A Life With Birds

Field Notes from a Hidden City: An Urban Nature Diary

BETWEEN LIGHT AND STORM

How We Live With Other Species

Esther Woolfson

GRANTA

Granta Publications, 12 Addison Avenue, London W11 4QR

First published in Great Britain by Granta Books, 2020

A CIP catalogue record for this book
is available from the British Library.

2 4 6 8 9 7 5 3 1

ISBN 9781 178378 279 6
ISBN 9781 178378 281 9

www.granta.com

Typeset in Garamond by M Rules

Printed and bound by CPI Group (UK) Ltd, Croydon, CR0 4YY

To Chicken,

Corvus frugilegus

Spring 1988–9 December 2019

Colleague, companion, friend.

For the animal shall not be measured by man. In a world older and more complete than ours, they move finished and complete, gifted with the extension of the senses we have lost or never attained, living by voices we shall never hear. They are not brethren, they are not underlings: they are other nations, caught with ourselves in the net of life and time, fellow prisoners of the splendour and travail of the earth.

HENRY BESTON, *The Outermost House: A Year of Life on the Great Beach of Cape Cod*

I would have liked, once in my lifetime, to communicate fully with an animal. It is an unattainable goal. It is almost painful for me to know that I will never be able to find out what the matter and structure of the universe is made of. This would have meant being able to talk to a bird. But this is a line that cannot be crossed. Crossing this line would be a great joy for me. If you could bring me a good fairy who would grant me one wish, this is the one I would choose.

CLAUDE LÉVI-STRAUSS

The history of man's efforts to subjugate nature is also the history of man's subjugation by man.

MAX HORKHEIMER

Contents

Beginnings

A late October afternoon. It's quiet. The blue light of dusk beyond the windows melts into early darkness. I'm in the company of others but I'm the only human being here. I'm walking from room to room, tidying, putting things in order, preparing for the evening when I notice a smirr of shadow passing over the surface of the kitchen floor. It's faint, just an impression before a glance, a small wisp of something, of blown feather, a dust-ball gusted in a draught. In these old houses, floors have weather of their own: breezes, cyclones, polar easterlies. I follow it closely until I see that it's walking, minutely but steadily across the desert expanse of floor, a spider so tiny that she freezes me where I stand, hyper-aware suddenly of my feet, of my own power, my murderous boots. This is a fellow inhabitant of my house, brought in by the cold, the incessant rain. In autumn, they all begin to look for shelter and for food: the house mice, the field mice who will arrive soon, the spiders, large and small, like this one. In summer, houseflies gather, wasps rattle in the window corners. All year, woodlice criss-cross the rooms with determined crustacean tread. There are too, the many creatures I can't see, ones too small for shadows.

This spider's clearly heading somewhere but I know I'll have to interrupt her journey in case, in doing other things, I forget she's there, step on her and end her life. Keeping my eye on her, I tear a page from a small notepad and bend to urge her onto it, her own brilliant yellow, magic-spider-carpet. Instead, she climbs onto my hand and walks about on my fingers for a while until I encourage her onto the paper so that I can carry her to safety, away from me, her only obvious danger. As a representative of my species, by comparison with this creature, I'm huge. As a member of my species, I carry the inescapable burden of the long, egregious history of human dealings with the lives of others. As an individual, I'm guilty by deed and by association. An almost-invisible spider faces me with myself.

As I hold this other living being on a scrap of paper, I know that there isn't exactly a relationship between her and me but a skein of connections which ties us both into the centre of the questions that have been occupying my mind for a long time, ones I ask now of a member of an arachnid genus and family I can't immediately identify (*Linyphiidae? Tegenaria?*). What are we doing here together? How, in the light of the hundreds of millions of years of our shared past should I behave towards you and others? They're questions anyone might ask themselves from time to time, anyone who lives curiously or anxiously as many of us do now, frequently with bewilderment and anger, often in despair. The first question is easier than the second. We can trace back to where our common origins lie, the progression from the earliest beginnings of life to where we are now, the point where one single species threatens the future of all the rest.

The second question is different, almost too much for one human and one spider, too heavy for the line that connects us both to those common origins, the one we can trace back through history and the stout, enduring certitude of the people who have argued and prayed and postulated through the centuries to ask what we're doing and why. It seems part of a long and continuous line which threads back through the chronology of all our lives, tying us to everything we have done and do, the questions strung along it like faceted beads, winking light and darkness.

I carry the spider carefully, one hand round the edge of the paper to stop her from falling. The two of us are so different, in size, components, physiology. We live differently, do things differently. We eat differently, grasp differently, see differently and, were we to kill, we would carry it out differently. For all that, I know us to be the same, sharing far more than just a long and parallel past. One of the things we share is life. That alone should be enough to bestow on us both a wild, determined equality, this minute scrap of living entity and me. The other thing we share is occupying our own designated place, determined by the species we belong to, on the planet on which we live.

As she heads off towards the edge of the yellow paper, all the exoskeleton and chelicerae, the spinnerets and many eyes and legs and the book lungs of her, I tip the paper slightly, holding it low so that she can resume her journey. In a moment, she is gone from the paper into her own shadowy, crepuscular realm behind the fridge, leaving me almost alone with all those thoughts and doubts we *Homo sapiens* have been considering since our long-ago brains sparked into fire and only sometimes light, thoughts relating to the many other species living on this earth, to size and power and rights in both past and future, to the consequences of our deeds, our ignorance or knowledge and on occasion, to the scale of our own personal culpability in the profound, lasting exploitation and destruction impelled by our species' needs and desires, its cruelty and greed. I try to imagine the concerns she's taking with her into that other universe behind the fridge. I could assume she feels relief in escape but that might be to imbue her with feelings she doesn't have. If I'm reluctant to endow her with humanlike feelings, it's not because I think she has no feelings or believe hers to be any less in strength or magnitude than my own. She might be more capable of deep emotions than any human but what exactly they are and how she experiences them, I don't know and never will.

Left alone amid the lights, phones, voices from the radio, all the crowded, overlapping spheres in which we're all obliged to find our uneasy way, I still move cautiously. If I had killed the spider accidentally, I'd have felt bereft, even if it would have been a brief bereavement. I've probably trodden unknowingly on dozens of similar creatures today.

Once, I wouldn't have thought very much about spiders or any other creatures or considered that they might be less than a human being. I loved and thought about the family pets I had as a child whose lives were close to mine but I wouldn't have asked myself why, or even if we were different, apart from in purely morphological ways. I never believed that humans occupy a higher position on any scale of life on earth than every other living being but if I had once considered the possibility, I probably would not have recognized the idea as part of an enduring inheritance of a particular culture and system of thought that has directed the attitude many humans hold towards everyone else who lives.

For the past three thousand or so years, we have nurtured a set of beliefs as unchanged as an insect waving immobile from its cage of amber, like some lost-cause corpse hovering in cryonic vitrification. These ideas have underpinned our behaviour, encouraged and allowed every deed and action in and towards the natural world. Over the vast territory, physical, temporal, cultural and intellectual, included in the term 'Western', we've clutched to ourselves with a rare and fervid assiduity a system of beliefs known as the *Scala Naturae*, or 'Great Chain of Being'. They derived from a fusion of biblical ideas, those of ancient Greek philosophers and later Christian theologians who assured us – speaking as they did with the highest divine authority – that being human in itself gave us ascendancy over every other species and that whatever we did to, or required of them, as individuals or entire species was justified because they were put on earth specifically for our benefit. Not only were we humans made in the image of God, we were the sole species given dominion over everyone else on earth, the only ones with souls.

If the astonishingly brilliant, inventive and thoughtful mind of

4

Aristotle provided a 'scientific' basis for a belief in a hierarchic scale of values of life on earth, it was the dazzling drama of Genesis which provided the background and explanation for it in a story that would entrance with its complex, illustrative beauty: the creation of a universe summoned, lit and populated within the seven days of its sequential, metaphoric cadences. The ideas engendered were to dominate half the world, as if, throughout history, almost every philosopher, savant or divine of any stature in Western thinking conspired against the interests of the rest of existing, breathing creation.

I might never have thought about any of this if I hadn't been impelled to by the random occurrences that dictate all our lives, the misfortune that causes birds to fall from trees, to become injured, abandoned, lost. At the beginning, when I first came by these birds, I knew very little about them but not knowing what else to do my family and I took them in. The feeding, care and housing of creatures acquired by accident or serendipity was simple enough and involved little more than consulting books and doing what seemed appropriate. Nothing told me that with these apparently chance acquisitions, I'd be challenged in every assumption I might have had about the lives and abilities of both humans and other creatures. If I had been concerned about my relative position on the branches of the Tree of Life, anxious to retain my privilege one tier below God and angels, I'd have felt the first swaying and tilting as the tree was blown by a mighty, excoriating wind. If I had ever believed humans to be the only ones to live profound and interconnected lives, I couldn't any more.

I don't remember now the first moment of realizing how much more the birds were than the very little sum of what seemed like their worryingly fragile parts, or that the rats my daughters Rebecca and Hannah kept were intelligent, complex beings but within a short time we knew that each creature possessed an indefatigable determinedness of self, a character, a place in their own world formed by geographical origin, long culture, social habits and cognitive ability. Observation told us that these creatures anticipate, enjoy, consider, grieve, love and hate. It was impossible not to appreciate that their actions were deliberate and their feelings as subtle, perceptible and as real as our own.

The timing was fortuitous – even a few years before, it would have been difficult to suggest that my family's observations had any basis in fact. The early years of our living with other species coincided with radical developments in studies of the cognitive behaviour and neuroanatomy of both humans and other species. Increasingly, news reports (many beginning with a would-be humorous reference to the term 'bird brains') spoke of freshly published research on the topic of avian, or other species' intelligence, and it seemed incredible only that it had taken quite so long for anyone to have noticed.

These revelations were exciting and seemed transformative. If we live in a world of sentience and consciousness, shouldn't the knowledge alter the entire basis of our relationship with other species? If these organisms are able to employ their statocysts, their chemoreceptors, mechanoreceptors, their neurons and glia, their nidopallia caudelaterale and hippocampi, their every organ and mechanism of sense, smell, direction and feeling towards living their lives much as we do ours, on what basis can we still consider ourselves superior?

The more I read and thought about what was being revealed by research and observation, the more the Scala Naturae and all the ideas that surround it came to seem the shabby, self-serving and self-destructive concepts they are. The construct appeared to represent the kind of ideas we should have examined, laughed at, and thrown away years ago, together with the anachronistic, scientifically incorrect language we still use: 'inferior', 'superior', 'higher', 'lower', 'evolved', 'less evolved', 'primitive', 'advanced'.

So extensive and damaging have been the effects humans have had on other species and on the fabric of the earth that they're described as our 'cutting down the tree of life'. Others have spoken of 'the sixth extinction' caused by massive species loss, climate change and the environmental degradation caused by human action. And yet, we still hold on to ideas of human perfectibility as if the greatest prize of all life and evolution is to be human, the end-point of some phylogenetic metempsychosis as we jostle for the nearest place to God. I'm not 'more advanced' than my rook, my crow or the spider on the floor. As a human, I'm not even among the distinguished roll of the most successful living species. I'm simply one small part of another small part of the life of the planet. The belief that humans are not only exceptional but superior to every other species is being shown increasingly to be absurd – it seems exceptional in itself that it has taken the imminent catastrophe of species loss and global warming to begin to demonstrate what we all might have known all along, that we're only part of a chain, not the sole and brutal wielder of it.

In writing this book, I've tried to answer, not least for myself, the questions I put to the spider, questions about our shared origins, about the development of the philosophical and religious ideas that have formed our views and how we've seen, described and portrayed other species. I've asked what has allowed us to keep and kill sentient creatures in vast numbers for food, clothing, sport or entertainment. I've asked how we reconcile the sufferings of others with benefits to our own lives and health, and what kind of relationships we have with the animals we live with most closely, how we choose those to whom we will be close, those we will defend or those we won't. I've asked too what moves us to love or hate or grieve beyond the boundaries of our species, how our behaviour and attitudes towards other species is related to our behaviour towards other humans.

The questions are slippery – they fall, break, scatter, send out splinters. If we're post-religious, secular, rational and scientific, why aren't we more prepared to change the way we think? In deposing the deities from the topmost branches of the Tree of Life, have we simply upgraded ourselves to their vacant seat-on-high? If we're the gods now, shouldn't we be better than we are?

Darkness. Just for this moment, the house with its thick stone walls feels protected against the weather but in the end nothing keeps it out, not the changes of the physical world, the effects of time, seasons, years. None of us is different. Together, we cling in whichever way we can to the surface of the earth.

I

Sharing a Planet

On the wall beside my desk, in a photo, the Berlin specimen of
Archaeopteryx sleeps her long, deep sleep under the glass of her frame.
This creature belongs to something more than time. I've put her where
I have only to raise my head to look at her in her bed of stone, at her
long bones bent and arching. Her three-clawed 'hands' beseech, her
wings open towards eternal, earthbound flight. Shadow impressions of
her long-gone feathers fan into the wide, veined arcs of ginkgo leaves
and butterflies' wings. She's beautiful. When she was found 150 years
ago, Archaeopteryx's fossil skeleton was lying perfect and complete in
the fine-grained Jurassic limestone, the Lagerstätte, which had held her
for 150 million years. She was called '*Der Urvogel*', 'the first bird', but
her given name is *Archaeopteryx lithographica* or 'Ancient Wings from
the Printing Stone', after the stone that entombs her, used since the late
eighteenth century for the making of lithographic plates. This bird was
the first of many things – the first Jurassic paravian theropod found,
the first to occupy her celebrated place in the record of palaeontology,
the first proof of a transition point between bird and reptile, the first
creature discovered with morphological features of both dinosaur

and bird, among them feathers, teeth *and* wings. She's recognizably related to the creatures I have come to know best. Despite her age, she is family. It's her angles and joints, the shape of her head, her neck, her feet. It's her very *birdness* that makes her so.

Behind me, pottering on the floor, is my own *Urvogel*, my own first bird, a rook, member of the genus *Corvus*, who lives out the days of her extreme old age. Her theropod dinosaur origins are all the clearer now as she ages, moults, her feathers becoming seasonally thinned to show again the terrifying balance between fragility and endurance. She's older than most of her kind because for three decades she's been protected from the dangers of the wild-bird world. She's thirty-one. Birds can be long-lived – a list of the ages of ringed wild birds I find records their endurance: *golden eagle: 32 years; avocet: 27 years 10 months; curlew: 31 years 6 months; common gull: 33 years 8 months; Robin: 19 years 4 months* (and this doesn't include the parrot family, the longest lived of all). These were the exceptions, the ones who by chance and fortune evaded the many dangers of predator and weather for longer than their kin.

The rook, whimsically named Chicken all those years ago, isn't my only bird. I have a crow called Ziki too and, for the moment at least, the blue-white dove who stands alone on the roof. She's watching too but I don't know what for – for her family to return perhaps, her companions, or her enemies. Until recently, there were a dozen birds in my dove house but when I went to feed them one morning a few weeks ago they were gone. Whoever raided their house had taken the single fledgling. I searched the garden and small lane behind the house, peered over the walls into my neighbours' gardens but there was nothing, no blood, no feathers, no prints. Most of the usual suspects, like cats or rats, leave evidence behind. The wire over their door, stoutly nailed in place for decades and regularly checked, had been ripped away from the lower right-hand corner and everyone had gone. My ancient brown dove and his daughter, the other elderly residents and the few visiting semi-wild birds sat along the roof gutter for a day or two and then they too were gone, too scared perhaps, ever to return.

The bird outside the window is the last of her line, her colours those of one of the earliest of the doves I kept, a sturdy, handsome bird, white with blue markings, a prolific and enthusiastic breeder, his devoted mate the first white dove I ever owned. She's the offspring of the offspring, although unto how many generations I've no idea. Before they scattered, some of the inhabitants of the dove house were well over twenty. I laughed, called it my care home for geriatric birds. It was always more sanctuary-hotel than proper, organized pigeon loft. These birds and I lived for years together in what seems now like ramshackle but easy harmony although it wasn't really – I've always been too anxious about them for anything like ease. They accepted food and shelter with calm, uncritical grace and I welcomed their presence, grateful for their fortitude, their fidelity to place and time.

Sometimes when I look out for my dove, she isn't there. I watch the downturned leaves on the small maple outside the dove house move and dance, dark red and pointed like the fingers of many hands frantically playing an invisible piano in an orchestra of blowing wind. Is she waiting on the roof for me to come out of the back door to lay her food out on the large flat stone beside the door? I've kept their house as it was, the platform where the birds went in and out, the wire-covered screen, mended now, the perches. Sometimes, she and the odd wild companion go in at dusk but I no longer go out as I've done every evening for decades to shut them in. I'm afraid that if I do, I may never let them out again. Now, from the corner of my eye, I see my solitary dove, the silvery flash of her wings as she descends to eat.

Doves are different from the other birds I've known – they don't form relationships with humans in the way corvids or psittacines do, with their relentless determination to *connect*. Doves pursue their lives with fierce, commensal grace. I was always thrilled by them, by the beauty of their flight, their weird enthusiasm for winter-ice bathing, by every earthly manifestation of their independent will.

I've always called them 'doves' because the first ones I had were white but over the years they were the ones most easily picked off by sparrowhawks, and in time wild doves, or pigeons, found their way into the dove house, birds of grey, lavender, slate, pearl. The designations are meaningless anyway – dove or pigeon, they're all *Columba livia*, descendants of wild rock doves. They evolved from the Jurassic to the Cretaceous and although the fossil evidence is slight, their family, the *Columbidae*, are probably birds of the Miocene, 20 million years or so old, birds of mountains, cities, stone.

Archaeopteryx too belongs to stone. By now, she's almost part of it, as if she grew from it, bent into it in the elegant, opisthotonic backward arching of many dinosaur fossils. Once, the pose was thought to be a manifestation of death agony but more probably was caused by post-mortem contraction of the ligaments, which once supported their considerable necks and heads. Ancient creatures aren't found only in stone but in amber too, the hardened, fossilized resin of coniferous trees. Insects, spiders, leaves, feathers and even entire small birds and lizards have been caught in amber's swift and unrelenting grip. There's a reconstruction I've seen of a fledgling discovered encased in amber, one of the most complete ever found, a mid-Cretacean enantiornithine from Myanmar, an avian group that died out 66 million years ago. Beak open, wings wide in upward flight, he's like any of the new small birds we watch for in spring and early summer, the ones just about to fly. In the resin that trapped him 100 million years ago, he glows, still perfect in his translucent, once-liquid gold.

This measure of time itself is difficult to figure, to take beyond the limits of today, the usual, straight-line, one-fixed-point-to-another concepts in which we're bound. The French philosopher Bruno Latour suggests that many people today, those he calls 'the moderns', understand time that passes, 'as if it were really abolishing the past behind it', like Attila, 'in whose footsteps no grass grows back'. The geologist Marcia Bjornerud thinks that we need to establish a new, more informed relationship to time, a new understanding which acknowledges a common past and future – a relationship she describes as 'timefulness' – in order for us to treat both the planet and its other inhabitants better. What she describes as 'environmental malefactions and existential malaise' stem she believes from our distorted view of our own place in the history of the natural world and our inability to see ourselves properly in relation to the past and the future.

Birds seem a way of understanding time. The ones I've known have given me a glimpse of the vastness of the past, reminding me that we're latecomers, evolved slowly into a world already old. They've changed my perception too of ordinary time. Every morning, I'd feed the doves and open the small door above their landing platform to let them out. Every evening after dusk, I'd shut them in. Without them, the rituals of morning and evening are unnecessary, the days no longer have the same rhythms. The light in the rooms has changed, no longer animated by the sudden patterns of brightness and shadow which lifted or fell according to their flight. In the garden, there's a layer of silence where the background sounds of their gossip and their grumbling used to be. It's odd and unsettling to be without the altering registers of their outrage or contentment.

Time lives in background sounds, the ones which accompany our lives, the messages the world gives out in sound waves, airwaves, voices heard and unheard, the purposeful dialogue of birds. Listening with an ultra-sensitive horn antenna at Bell Lab in New Jersey in 1964 to try to detect and analyse radio signals from between galaxies, the physicists Robert W. Wilson and Arno A. Penzias heard through it a faint, unidentifiable hiss. At first they thought that it was faulty equipment or an anomaly caused by droppings left by the pigeons who were nesting inside the antenna but after they'd cleaned the equipment and removed the pigeons, they understood that what they were hearing was background radiation, the echo through time from the first moments of the universe's beginning, 'a faint glow of light that fills the universe' as the physicist Erik M. Leitch described it, 'the oldest light we can see', the 'afterglow' of the Big Bang. The finding of Cosmic Microwave Background, for which Wilson and Penzias were awarded the Nobel Prize for Physics in 1978, was to be one of the foundations of modern cosmology, with one of those who further expanded the understanding of it, James Peebles, being awarded the Nobel Prize for Physics in 2019. (The pigeons who were nesting in the antennae received no such honour. In an interview, Penzias talked about them being sent away to a pigeon fancier but how, in the way of pigeons, they swiftly returned. Then, Penzias said, they were shot, small recognition indeed for their irreplaceable role in the discovery.)

It was when I began to write about the origins of birds, tracing them back through the layers of palaeontological time to understand the origins of my rook that I went back to the moment when my path towards being what I am branched off from the path of birds some 280 million years ago, and felt a pang of retrospective regret, wondering pointlessly enough if I might not have preferred to be someone or something else, or at least wishing that I might have kept the pleasant, intelligent company of corvids through the long span of years.

I try to look for the fine connections. Who knows how life began, an immutable glint of whatever may have begun from one or other of the long processes still not entirely understood? There are so many ways suggested, from the organization of mineral crystals in clay, from the protected environment under many feet of the ice that sheltered organic compounds from the dangers of ultraviolet light and cosmic impact, or from deep submarine hydrothermal vents – hot, bacteria- and chemical-laden springs that bubble up from volcanic fissures in the sea floor. Discovered first in 1977 by scientists working in the Galapagos, these vents form long, deep 'chimneys' through which rise clouds of chemicals in temperatures of up to 700° F – 'black smokers' carry fine particles of iron sulphide and 'white smokers' barium, silicone and calcium. In 2015, a team from Uppsala University led by Thijs Ettema, discovered a microbe which they believe demonstrates that eukaryotes emerged from apparently simple single-celled archaea, altering the long-accepted view of cellular life, which suggested it was comprised of 'three domains' – bacteria, archaea and eukarya – while the question of the relationship between archaea and eukaryotes hasn't yet been settled by the findings, Ettema believes that they'll contribute to the 'uncovering of our microbial ancestry'. The work was done on samples gathered from a hydrothermal vent 2,300 metres deep in the Arctic Ocean between Greenland and Norway. The site, named 'Loki's Castle' after the eponymous god, is a 'black smoker', which seems appropriate somehow, as if it might have exploded, billowed from the pages of the great work of Nordic mythology the *Edda* and a fitting home surely for microbes named heroically *Thorarchaeota, Lokiarchaeota, Odinarchaeota, Haimdallarcheaota* in the super phylum 'Asgard' in tribute to those mighty gods Thor, Loki, Odin and Haimdall.

In 1952, in a now famous experiment, research chemist Stanley Miller, working at the University of Chicago under the supervision of Harold Urey, winner of the 1934 Nobel Prize for Chemistry, attempted to demonstrate that the basic chemical components of life on earth could be synthesized by replicating the conditions of early Earth. Earlier, the Soviet scientist A. L. Oparin and the Briton J. S. Haldane had separately developed 'primordial soup' theories which suggest that life originated from molecules interacting with the atmosphere to create simple organic compounds, the process of 'abiogenesis' by which life arises from non-living, abiotic compounds. (The word 'abiogenesis' itself is one to drive creationists to frenzies of outrage and to the immediate and urgent expenditure of large amounts of money in the construction of websites promoting the refutation, by whatever means, the idea that life began in any way other than by the direct and immediate action of the hand of God some six thousand years ago.) The Miller–Urey experiment, designed to prove Oparin and Haldane's work, was conducted using simple equipment, glass flasks of water, methane, ammonia and hydrogen and electrodes to simulate lightning in an approximation of the conditions of early Earth. The resulting development of a number of amino acids required for the formation of proteins in living cells contributed towards this proof and while the experiment has been criticized, examined, re-examined, refuted, re-evaluated and re-validated (as well as being overtaken by subsequent findings about the atmosphere of early Earth) it remains seminal in its demonstration of the synthesizing of the basic components of life.

I trail the line backwards. All living beings stem from a single-cell organism resembling a bacterium, anaerobic and autotrophic – not breathing oxygen, feeding itself from synthesized inorganic material – designated LUCA, the Last Universal Common Ancestor. LUCA lived 4 billion years ago, and while it's not yet known exactly where it first occurred, it probably originated in that hot, deep water bubbling from under the sea floor. We may share a long history but it's difficult to feel any sense of identification or family connection with LUCA, our possible microbial ancestor lurking for billions of years in its bubbling, smoking, chemical Jacuzzi. Only its name and the word 'ancestor' inspire a mildly Confucian sense of respect, a faint but indelible tracing from there to here, a line of evolution so slow as to crack the credulity of our instant age.

As a past and a beginning, there is the long, slow burning of the Hadean, 4.6 billion years ago, a boiling in gas and dust and a slow cooling, turning by 4 billion years ago to Archean rock and the memory in our oceans of millions of years of falling rain, of water dissolving, minerals combining in processes by which the formation of the earliest forms of life could evolve. The evidence appears to show that they did first as bacterial micro-fossils, cyanobacteria, the stromatolites and biogenic graphite found in Greenland and the detrital zircon grains of Western Australia, estimated to be some 4 billion years old.

Oxygen may have allowed the growth of life on earth. The eukaryotes, the aquatic choanoflagellates, sponges, ctenophores, the echinoderms – starfish and brittle stars, cnidarian sea-pens and sea-lilies, the Ediacaran fauna such as the frond-like *Charnia* of the pre-Cambrian 635–543 million years ago, were joined by trilobites, molluscs, corals, crinoids, brachiopods, nautiloids, the earliest fish, the first chordates. Life flourished during the 'Cambrian Explosion' some 543 million years ago when most animals found in the fossil record appeared, diversified, leaving their remains behind in the rich seams of the Burgess Shale in Canada and Chengjiang beds of China, writing the details of their physical forms into sediment and stone, the first true vertebrates, conodonts, euthycarcinoids, possibly the link between insects and crustaceans.

Along with my *Ur* bird on the wall, I've kept a replica trilobite in my desk drawer for years, a Silurian *Sphaerexochus britannicus*, 400-million-year-old relative of the woodlice who march across my floors. He keeps close company with the other drawer resident, a small replica *Diplomystus*, an Eocene fish.

The brilliant, endlessly inventive, often contentious evolutionary biologist Stephen Jay Gould believed that, were we able to rerun the Cambrian explosion, *Homo sapiens* as a species might not have happened or might have been something else entirely, describing our existence in such a large universe as being: 'a wildly improbable event well within the realm of contingency'. This thought was, he said, a source of both freedom and 'consequent moral responsibility'.

In 2014, a Late Carboniferous fossil arachnid, 305 million years old, was found in the sediments of Montceau-les-Mines in the Saône-et-Loire and described in 2016 under the title 'Almost a Spider', because the creature appears to be a link between early arachnids and modern spiders in being able to produce silk while not having the spinnerets required to form it into webs. She was given the name *Idmonarachne brasieri*, her genus named for the mythological Idmon the dyer and his daughter Arachne who famously was turned into a spider, and her species for the palaeontologist Martin Brasier who tragically was killed in a car crash at the end of 2014.

During the Early Ordovician Period at around 479 million years ago the ancestors of insects originated. Eighty million or so years later, insects flew. Corals appeared and nautiloids. Around 450 million ago, non-vascular plants such as bryophytes evolved, the first vascular plant *Cooksonia* developing some 30 million years later.

Around 400 or so million years ago, aquatic animals were becoming land animals. Non-vascular land plants grew, forests evolved and ferns and fungi grew in the forest shade. Spiders evolved from aquatic forms – the first fossil evidence of true arachnids dates from 380 million years ago. The first four-legged animals, tetrapods, evolved from intermediate creatures such as *Tiktaalik roseae*, a 'fishapod', which had features of both fish and amphibian, precursor to much later forms, amphibian, reptile, bird and mammal. The tetrapods split, one branch becoming the amphibians which would develop into lizards that would in time become dinosaurs; the other branch splitting further into sauropsids and synapsids, the former becoming modern reptiles and birds, the latter evolving into mammals. One group of synapsids, the pelycosaurs, thrived but were replaced by their close relatives, the therapsids, creatures rather like shrews which evolved during the Permian Period 298 million years ago, one group of which, the cynodonts, were to evolve into the first mammals, frogs, turtles and crocodyliformes, ancestors of modern crocodiles. Silk-producing spiders evolved, rain forests, the coal forests of the Carboniferous flourished. After the huge Permian extinction, those sauropsids which were not destroyed evolved into dinosaurs. Reptile groups became aquatic, developing into marine dinosaurs: the plesiosaurs, ichthyosaurs and mosasaurs.

And then, the early birds arrived – 168 million years ago,

Epidexipteryx hui, a flightless dinosaur with what seem like the first-discovered display of tail feathers, lived, possibly an early forerunner of modern birds and, 18 million or so years later, *Archaeopteryx* appeared.

Plants and flowers evolved, and the insects who would feed on them. *Archaefructus*, a plant with carpels but no flowers evolved, followed by flowering plants; the first to be found as a fossil was a flower resembling a magnolia. Kalligrammatid lacewings who, from fossil evidence look remarkably like butterflies, were creatures of the Jurassic and Cretaceous.

In the Early Cretaceous Period, 145–65.5 million years ago, placental mammals split from marsupials (although there is evidence that this may have happened earlier), splitting further around 105–85 million years ago into four groups, one that would evolve into ungulates, another into whales, bats, primates and rodents, the third into anteaters and armadillos, the fourth into elephants, aardvarks, hyraxes and other species.

There were extinctions – from the ice-age event of the Ordovician, 443 million years ago when 85 per cent of marine life was destroyed, the long, slow 'climate-change' of the Devonian Period 360 million years ago, the Permian–Triassic 250 million years ago – often described as 'the big one' when 80–95 per cent of all marine species were destroyed – and the Triassic–Jurassic, 200 million years ago, with a loss of 50 per cent of marine invertebrates and 80 per cent of land quadrupeds.

By 100 million years ago, the famous dinosaurs of the Cretaceous Period, the tyrannosaurs, *Triceratops*, the *Iguanodons*, the *Troodons*, had reached their zenith and were about to be overtaken by the Cretaceous–Tertiary, the K–T, also known as the K–Pg, or Cretaceous–Palaeogene extinction event of 65 million years ago, which destroyed many other species: most dinosaurs, most tropical marine life and many plant species. The mechanisms by which some species weathered this event aren't completely understood but small mammals, some birds, crocodiles, horseshoe crabs and platypuses were among those who survived. The survival of some bird species has been attributed to their large brain capacity and consequent adaptability, as well as their fortuitous ability to fly.

(Often, the history of the earth seems like a bad film script, too unlikely, too fantastic in a ferment of volcanic eruptions, asteroids crashing, climates flipping from fire to ice. *Five mass extinctions? Wouldn't two be enough? The destruction of 90 per cent of species, the destruction of most dinosaurs – really?*)

Time scrolls on, life scrolls on, returns, thrives, is lost, returns and thrives again. In the Cenozoic Era after the K–Pg, there was rapid mammalian radiation – during the Palaeogene Period, 65–24 million years ago, the primate groups haplorhines and strepsirhines split, the latter to become lemurs and aye-aye, the former monkeys, apes and humans.

New World monkeys diverged from the other primates 40 million years ago to colonize the area that is now South America. Around 15 million years later, apes diverged from Old World monkeys, followed by what might seem like a series of primate-family feuds, a gibbon–ape divergence 18 million years ago, an ape–orangutan divergence 4 million years later, a gorilla split 7 million years later and a chimpanzee–human ancestor divergence about the same time. *Pierolapithecus catalaunicus*, the middle Miocene hominid, possibly the last common ancestor of gorillas, chimpanzees and humans lived 11.9 million years ago; *Orrorin tugenensis* 5.8 million years ago, *Ardipithecus ramidus* 5 million years ago, succeeded by members of the genus *Australopithecus* who lived between 4.2 and 1.1 million years ago.

By 2.5 million years ago, hominids were using stone tools; 600,000 years ago *Homo heidelbergensis* inhabited parts of southern and eastern Africa and Europe. It was thought once that *Homo sapiens* evolved in a line from *Homo heidelbergensis*, the last link between *Homo sapiens* and Neanderthals, but more recent discoveries – of *Homo floresiensis*, found in 2003 in an Indonesian cave, *Homo naledi* discovered in South Africa in 2015 and *Homo luzonensis* found in Luzon, Philippines first in 2007, with further finds made in 2019, of Denisovans in Siberia and Tibet – have changed the view of past relationships, place and time. The expansion of DNA studies promise greater understanding of the slight evidence they left behind.

And then, with us, everything seems to slow, to concentrate down, as if time collapses when we're around, billions to millions, millions to thousands and the line runs out. With us, it often seems, time is the moment. 'The moderns . . .' Bruno Latour writes, feel such discontinuity from the past that they believe, 'nothing of the past survives in them – nothing of the past ought to survive in them', although of course it does.

There's a place I drive to just beyond the city and the suburbs, at the edges of the farmlands. I turn off onto a small road that winds through low hills, past a goose pond, empty of geese for most of the year and covered by a slick of growing green, past the handwritten signs that in spring warn of young lambs on the road. Half a mile on, a small bay of stones is set among grass and trees. Sometimes I drive past it and sometimes I stop to park opposite the information sign put up by a historic organization responsible for its upkeep. This is the site of a Bronze Age stone circle. Its existence is breathtaking but not unusual around here; Neolithic stones can be glimpsed from the road, entire, majestic circles standing or recumbent, tall and towering solitary in fields. Individual stones lie, fallen, toppled, half buried, as if they're being hauled back into the earth, as if they're there to remind us whence, in substance and in time, we came. Writing of rocks and stones, the geologist David Leveson says: 'Together with the mass from which they have been broken . . . they state the framework of my existence, permit it, define its possibilities, outline its limitations.' Here, even the far past seems so near that if I turn my head I hear the sound of the clicking of its abacus, catch the steady movement of its finger, tracing. I think of it as the companion who carries every story, remembers everything we've been and are.

By now, I've come to an easy familiarity with the stones. The circle seems like a small, enduring gateway into the past of the world, into the 4,000 years since this ring was constructed, since these eight granite boulders were placed carefully on this valley land scoured first by fires of willow; 1,500 years may have elapsed before the inner ring of cists was added, these very cairns where cremated bones were found in ash of oak and hazel. Now, both stones and cairns lie in peaceful, rural quietude, looking as if they were constructed yesterday, neat and ordered, no visible bones, no remains of any sort, nothing to speak of the lives lived here, thousands of years of human lives. The stones of the circle aren't the towering kind – only one of them is taller by a few inches than I am, which at five feet isn't tall. The only indication of human involvement is their careful, measured ordering. The lives, the weight of the people who placed them here seem airy and light; their bones, their homes, every possession and every other creation is gone, merged into the acid of the land. All we can do is try to construct the life they had and who they might have been from what's left behind for picking through and analysing, reassembling half-pictures from broken fragments, from shell and ash and bone, not knowing for certain if anything we deduce or imagine is right, or wrong.

In her book *A Land* the distinguished archaeologist Jacquetta Hawkes suggests that people living near Bronze Age circles and menhirs believed the stones to be people transformed into granite and limestone. But as I stand beside them, the stones seem less former humans than small lithic voices whispering. Remember us.

Remember us. Stone has its own life too, its own history, a backdrop to our own. Everything connects us, leads us back. Who can really understand aeons and eras or create pictures from the names Hadean, Archean, Proterozoic, from long darkness, from gas and fire and the wild celestial bombardments during the passage and events of the 4.567 billion years since Earth was formed. In his 1795 book *Theory of the Earth* the Scottish geologist James Hutton wrote of his studies into rocks and the processes of their formation and stratification – what he described as 'materials furnished from the ruins of former continents'. In a remarkably succinct evaluation of his work on 'the great geological cycle' he concluded, 'The result, therefore, of this physical inquiry is, that we find no vestige of a beginning, – no prospect of an end.'

Often, I'm reluctant to turn away to leave. There is a stillness, a sense of calm in this place of stones. The thought of the lives of those who placed them and lived among them brings me a brief and welcome sense of distance from the day-to-day world and the events of the time when I'm writing this, days of political upheaval, random violence and change. They're probably not very different from the now-forgotten turbulence of those former lives. The stones remind me of the evanescence of the world's drama, the *Sturm und Drang*. I recall a quotation from that archetypal *Sturm und Drang* man, Schiller: 'On the mountains, there is freedom. The world is perfect everywhere, save where man comes with his torment.'

Years ago, when I was a student, I was stopped in an East Jerusalem street by a very old man who asked me in German for directions to the Augusta Victoria Hospital. It wasn't far away and I offered to walk there with him. He took my arm and together, using words from whichever common languages we could find, we exchanged a typical conversation for the time and place as we walked through the old streets to the gates of the hospital: 'Where are you from?', 'How did you find your way here?' As he said goodbye, he urged on me the book he was carrying. I tried to refuse but he insisted and walked away before I could give it back. It was an old copy in Gothic script of the works of Schiller, which I kept until it was burned with all my other books in a house fire a few years later. I didn't understand much of it but as long as I had it, its old blue cover whispered to me of its own framework of time, of its former owner's life and journeying, of human discourse and experiences I would never be able to imagine.

No one knows why the people who brought these stones here placed them as they did. Were they chosen as sites from where they could best watch the movement of the planets? Were they markers of power? Or simply places of ritual or burial? By comparison with the more majestic, towering stone circles of Callanish or Stonehenge, this one is almost homely in its modest scale, an unintimidating, peaceful place for the dead to lie. The land has changed since these placers of stones and buriers of the dead lived here. It has been made and unmade by humans and weather and years. It would have been bogland then, like much of the coastal land, low and damp, surrounded by dense tree cover. A few miles away there are the shadow traces of a Neolithic building, now almost imperceptible amid the lush, deep greens of northern farmlands. Once, it was a rectangular wooden structure with stout corner posts, probably a communal house built in a clearing among trees but now everything has gone except the revenant image, the ghost traces of that ancient life.

Excavations in a field nearby uncovered twelve pits dug 10,000 years ago which were identified as a lunar calendar, one of the oldest ever found. It lies still in perfect alignment with the phases of the moon and midwinter solstice, in use it seems for 6,000 years until the place was abandoned for reasons we'll never know. A sense of the lives of these people reaches me as a presence not quite there, but not quite extinguished. I reach out and touch the stones of rough, red granite. Like all stones, they induce reverie. They possess the power to bring us to the edge of the unimaginable. They hold tight to our uneasy dreams of deep time and remind us of their violent pasts, of volcanoes erupting, of their burning and melting, solidifying and freezing throughout the long, slow, ancient processes of impermanence and change. Against them, humans are no more than faint lines on those stark graphics of evolutionary time that show us so clearly how brief has been the tenure of our life on this planet and how much briefer the future for all of us might be. A speck of dust and *You are HERE*, all but invisible, looking back at the magnitude of what came before.

I drive on, up the slope past the stand of trees where the kites circle, up to the high point where the valley opens towards the mountains, fading blue and lilac towards the horizon. On some afternoons, the valley's bright with sunlit greens, on others, lost in a winter haze of pearl and grey dissolving as the land rises to the north, snow-covered into cloud and sky.

Smoke lifts and spreads from a farm bonfire on the hill, dissipates into the cool spring wind. In early summer, the sharp yellow of broom lights the road edges and hillsides, airy fringes of cow-parsley mist the boundaries between road and field. On some days, a single wind turbine turns steadily above a line of trees in stoic rhythm, as if telling us something we always should have known.

The wind carries faint scents of smoke, farmyard, broom and grass. Was there a time before the last hundred and fifty years when we thought in the way we do now of what we breathed, when we wondered about the composition of the air? I look at charts of atmospheric gases that trace the rise and fall of levels from Archaean times: methane, nitrogen, oxygen, fluorine, chlorine. They note the formation of the ultraviolet absorbing layer of ozone, which would protect the surface of the planet from the dangerous effects of the sun's UV rays. I find one that bears a legend down its side, a wavering line of changes in the history of the earth's atmosphere: *warm, cool, warm, cool, warm, cool*, it reads, telling of millions of years of fluctuating temperatures, Archaean to Proterozoic and further, to the time when life on earth could breathe, and beyond.

Warm, cool, warm, cool, warm, cool. It's a line that charts the prolonged, icy progress from 2.58 million years ago, and wavers on until today in the Quaternary glaciation – through 'glacials', 'interglacials', 'stadials', 'interstadials' and 'oscillations' – as Earth responds to changes in levels of carbon dioxide and alterations of its orbit affecting the way sunlight falls. For millions of years, ice froze and melted, spread over some parts of earth but not others in a complex pattern of periods of warming and cooling until the period of most severe cold, the Last Glacial Maximum, which ended some 20,000 years ago as the ice withdrew and the earth grew steadily warmer. The process was interrupted briefly 12,900 years ago by a sudden return to ice, the Younger Dryas, which is named after a small, white Alpine flower whose pollen was found in abundance in ice-core samples from the period. This cooling is thought to have happened as a result of the effects of the melting of ice sheets which, in their dilution of the saline content of the North Atlantic, changed the thermohaline circulation of the oceanic system, the force which drives currents, according to their temperature and salinity, a major determinant in climate alteration. The Younger Dryas itself ended abruptly 11,500 years ago, probably because of the reduction in the volume of water from melting ice. Temperatures rose again, some in the northern hemisphere rising by ten degrees within a decade, a major influence for future cultural and social change.

Behind the hill, the blades of the wind turbine turn. Our life here on earth has been so small in time, so great in impact. I think of the composition of earthly gases, 78.08 per cent nitrogen, 20.94 per cent oxygen, 0.934 per cent argon, 0.039445 per cent carbon dioxide, 0.0001818 per cent neon. It's carbon dioxide we know and watch. For most of the Holocene, the epoch in which we live, more commonly now called the Anthropocene, there were 275 parts per million (ppm) of carbon dioxide in the atmosphere. From the beginning of the Industrial Revolution levels began to rise until by 1960 there were 315 ppm and although the numbers seem small they're not. In 2014, Bruno Latour wrote of his shocked reaction to a report he'd read on the rise in CO_2 levels, wondering how we can be part of this long history, have played such a vast part in it and yet come to the realization of it so late, with such lack of ability to mend it. 'How are we supposed to react when faced with a piece of news like this?' he asked. Ten or so years ago there were 380 ppm of carbon dioxide in the atmosphere. At the time of Latour's writing the level was 399.29. As I write this, there are 412.

There are days here when the sky dazzles white and wide and high and others when it deepens slowly from chalk white to slate, when the margins of ash- and charcoal-coloured clouds flame suddenly with filaments of light. Red kites hunt above me, sometimes a single bird but often two, three or more, fork-tailed swirls of russet resting lightly on the air and it seems that there may be nothing as graceful as their turning, as their soaring into sunlight, their feathers changed in an instant to flashes of copper taffeta, light as fragments of fine silk blown by the wind. It's impossible to forget that they're birds who, with many others, were hunted to eradication from this area. Not long reintroduced, now they're here all year, dependable, for the moment in their slow drift, a point of constancy in an avian world of movement and migration.

Beside and above the road is the ancestral territory of rook and crow and jackdaw, a formidable presence visible everywhere in their rising and their circling, their calling from the denseness of the tall trees above the road. It's only a few miles from where, as a small fledgling, my own rook Chicken was found. They scatter over the fields in their feeding, rise to cast fine grey nets of roosting flight across the skies at dusk. Here, early spring is always frantic with transportation, construction, guarding; late spring is loud with demanding, an insistent symphony of yelling, muted in a moment to the gobble and gargle of young corvids being fed. In summer, swifts pierce the air with the lightning of their speed, their sweet and vibrant shrieking. Then, there's the annual autumn melancholy of the departure of these strange, liminal creatures. One or two are seen to leave later than the others, flickering solitary over the harvested fields. I always feel a sense of desolation, perhaps because they're going and I'm not, or perhaps because there isn't one of us who can be as sure as we once were that they'll return next spring.

In the years I've been passing this way, autumns have turned stormier, winters wilder, rain-laden, flood-bearing. Often now, I have to turn back because the farm road in front of me will have become lost into water, which by deep winter will have frozen into ridged and treacherous slabs of ice. The fields flood too and on the margins of expanses of newly formed field-ponds and grass-lakes, a few gulls always stand, calmly imperturbable in a misted landscape of chill and sparkling grisaille.

Close by to my right, there is a farm, its wire fence, sheds, caravan, pieces of random equipment lying in the common agrarian homage to rust, corrugated iron and old timber, the detritus and necessities of all farms everywhere. Ahead, beyond the stones, fields, low hills, the houses of the distant town. From across a wire fence, a few cows watch me with their sweet, mild air of benign detachment. On fine days, the scene's gentle. In summer, freshly moulted corvid feathers protrude from the grass, small spears of pointed black hurled from above by a determined, feathered army. The wind blows, seasonally fierce or douce or bitter through the high pine branches and the leaves of the rowan beyond the stones.

I turn from them and walk back across the grass. I don't see anyone else. The faint sounds of travel, engines, tractor, car and plane blended to the ubiquitous, signature buzz of humanity. At my back, a city, unseen. Ahead of me, road, fields, small towns, villages, castles, mountains. Under my feet, the steps of history, old landscapes, words passed down, things picked up, shells, stones, the remnants of wood whittled into points. Nearby, there's a small stone farmhouse. As I pass, it seems to stand as an almost-monument, a building that deserves to be marked along the scale of the processes of human life. A farm is a symbol of change, a startling innovation from our previous lives of stone, Palaeolithic to Mesolithic to Neolithic as we became who we are.

In the Lower to Middle Palaeolithic, we hunted, wandered and ranged as hunter-gatherers before we became complex and sedentary hunter-gatherers, collecting in groups, building, settling for most of the year, storing food. Through the Mesolithic into the Neolithic, 12,000 or so years ago, we changed how and where we lived in response to changes in the climate and the availability of food. At different times in different parts of the world, food acquisition from hunting and foraging was supplanted by food production and agriculture. Many people, including the pioneering agronomist and plant conservationist Jack Harlen have asked 'Why farm?' What provoked and encouraged this vast change from one way of life to another almost certainly more arduous way of life? There are too many variables – meteorological, economic, demographic and social for a single answer. Population pressures that encouraged disputes over land tenure, a lessening of adherence to a nomadic way of life, a scarcity of large animals possibly – or possibly not – caused by Palaeolithic hunters or climatic changes, as well as greater expertise in animal domestication and plant cultivation may all have contributed.

But we now had both opportunity and time to become something other than we had been. We now had the circumstances to make, to learn, to think. When I think about these human discoveries and innovations, I ask myself, could you have done that? Could you have invented anything? Could you have figured out how to make rope, needles, a net? Could you have made a basket or a harpoon? Could you have discovered how to make the beads to form a necklace? Could you have domesticated a pig, a sheep, a goat? (My answers wouldn't be promising for humanity.) With the advent of farming, everything changed, the future of the planet as well as the prospects for the reach and ambition of humanity. The archaeologist Ofer Bar-Yosef has described the transition to agriculture as 'a major "point of no return" in human evolution'.

As I cross the grass, there are no traces of how this place would have been, or at least none that I can see. I know people who can read the minutiae of landscape, who are able to strip layers of years away with their eyes, see the past by reading the stone of walls, contours, hollows, like ancient seers or visionaries, interpreting shadows as if they're magic, apotropaic signs.

The circle of stones allows me to imagine the people who arranged them, those who with utmost care, after minute observation, charted this measure of the cosmos, or perhaps simply and reverentially buried their dead. They would have lived their lives among trees, in the dark denseness of forests of birch, oak, elm, rowan, poplar, cherry. There would have been willow and hazel, juniper and myrtle. Even writing these names brings me the scent of plants and leaves in warmth and dampness, the smell of earth. The sounds they heard would have been the same but more, and less: wind through trees, rain, the voices of birds, animals and children, the hazed polyphony of insects, the road hum gone, the tractor engine, the distant helicopter, the unseen plane.

There would have been elk here and aurochs – distant ancestors of the watching cows – brown bears, wolf, wild boar and lynx, all now gone. There would have been red and roe deer, fox, badger, beaver and otter, wood mice – *Apodemus sylvaticus* – the stoats, polecats, weasels. There would have been wild cats – *Felis silvestris silvestris* – pine martens, hedgehogs and birds, capercaillie and lapwing, crossbills, doves, corvids, raptors. In the time since, some species have been hunted to the point of extermination and some have become extinct. Most species still here are fewer in number, some still hunted, some almost extinct, some just holding on.

Above me, birds hover, watching, flying. They forage purposefully in fields, trail in white flurries behind tractors as they move over the surface of furrowed ground, pause meditatively on fence posts and phone wires. I think of my missing doves and how they saw the world differently. Every day at dusk, I used to wonder how they found their way home, marvelling at their precision and timing. Once, I thought

I knew but now all the long-accepted theories about magnetism and particles of iron in their beaks have been questioned or proved wrong. It may be by scent or sight, by the action of the magnetically active protein cryptochrome in their eyes, by the sighting of the North Star or all of them together. We've long connected birds and stars in constellation's names, *Cygnus*, *Corvus*, *Aquila*, *Grus*, *Tucana*, *Pavo*, in their flight across the skies. In Edwin Morgan's poem 'The Archaeopteryx's Song' his almost-bird longs to be free from his imprisonment in stone, to soar lightly skywards towards his future, avian destiny.

On winter afternoons, I'd watch the doves going home, their hurrying, feathered figures of blue and white stepping through the arch of the doo'cot. The birds of the garden would have gathered themselves in the invisibility of branch and ivy. In the wild, my rook and crow would have been roosting. I think of them each time I watch twilight-driven clouds of rooks, crows and jackdaws flying like grey smoke over a freezing cobalt sky. Sometimes when I went to shut the doves in on Saturday evenings, if the sky was clear I'd watch for the three stars in the sky which designate the end of the Sabbath. I'd watch for the green-pink glow of the Northern Lights, the aurora borealis, for blue moons and blood moons, eclipses and meteor showers according to the seasons, as people everywhere have always done, observing, noting, calculating, reaching conclusions that were sometimes right and sometimes wrong. Early accounts are moving in their fear and wonderment – among the first written astronomical observations inscribed 3,500 years ago on the ox scapula and turtle shell of Chinese Shang dynasty oracle bones someone scratched: 'The sun was eclipsed in the evening. Is it good?', 'The next day was foggy, 3 flames ate the sun and there were big stars', 'Will there be a disaster today?', 'On the 7th day of the month, a great new star appeared in company with Antares.' I watch the sky but know I'll never feel the power of it as they do, these birds who have learned their orientations from stars and moon.

This morning when I went into the study, a faint white mark was impressed against the window, a perfect image of the pigeon or dove who must have flown against the glass in fright. When it happens, the feather dust from their wings leaves a ghostly imprint, invisible in some lights, clear in others. The wings always remind me of Paul Klee's *Angelus Novus*, an ink-transfer drawing of a funny little angel with fearful eyes, paper-quill curls and lifted wings. Drawn in 1920, it was bought the following year by the German philosopher Walter Benjamin who wrote about it in the ninth of his 'Theses on the Philosophy of History'. In it, he described the angel as 'the Angel of History', a figure he saw as emblematic of the failure of past certainties and the unpredictability of the future. For Benjamin at least, the angel might have been a portent. The drawing, said to be his most prized possession, was left behind when he fled from country to country seeking the safety he never found. He wrote of the angel being caught in a 'storm blowing from Paradise'. The storm, he said, 'is what we call progress'.

I return to the car and drive on, past the field corner where the cows like to gather to observe the passing world. After the piercing yellow of broom fades in high summer, rosebay willow herb takes over to bloom tall in purple-pink sentinel walls of flowers. I drive up the hill to where the fields open across the valley, past the trees where red kites hunt. Once or twice on this stretch of road, I've stopped to nudge a lamb back off the road and for a long time afterwards seem accompanied by that singular, inimitable scent of sheep. As I continue on my way, the all-knowing companion's still with me, whispering to me of what we've done and what we haven't done and why, asking me questions whose answers I only wish I knew. Above me, a circle of feathered, flying dinosaurs hover, as if pinned against the sky.

2

A Thing Apart

It's evening. I walk through the house switching on lamps. In his room off the kitchen, the crow tears paper recreationally, his radio playing Scarlatti. In the hall, Chicken the rook perches under the chair as she does increasingly now in these days of her old age. The chair moves slightly as she shifts in a series of small thuds for which I listen day and night to reassure myself that she's still OK. It rocks gently, wood against wood like the ticking of a slow clock. I leave a lamp on for her all the time now because darkness seems to frighten her.

True darkness is rare now. The darkness of a city night is false, deceptive darkness, interrupted by light, all the glowing, ubiquitous reminders of our human presence. Even outside cities it's difficult to escape from light, from the distant city glow. There are degrees of darkness, from the depth of winter darkness to the long northern, half-light nights of summer. We use the metaphor of darkness for the past, the veil, the arras separating us from who we might have been.

The first time I experienced true darkness, it felt chthonic, the darkness of numinosity, of mortality itself. That summer, I was just married and had gone with my husband David to where his parents lived in rural Zambia. The only electricity in the district was supplied by generator to the hospital where my father-in-law worked and so the alchemy of paraffin lamps was the sole way of transforming those few quick, tenebrous moments into fire and light. Walking out into evening only a few metres from the tilley-lamp-lit house, we'd melt into darkness, inhaling it instead of air. Switching on our torches in that dense opacity of lightlessness sparkled it instantly to terrestrial starlight. Glints from many watching eyes reminded us that both place and night belonged to a world beyond ourselves. One night, we camped on the high escarpment above the Luangwa Valley and watched the glassy clearness of day cloud towards dusk into a thickening glow of topaz and lilac light. Then darkness streamed down to enclose the world.

A few years before that, David and his siblings were wandering in the quiet, sparsely populated countryside miles from their home as they'd done all their lives, unsupervised even when small in the territory of lynx, civet and baboon, of hippo, aardvark and hyrax, hornbill and honeyguide. Climbing among an outcrop of stony hillocks – more rock piles than hills, they noticed the entrance to a small earth-floored cave. The moment they went in, they knew that the distinctive ochre and black marks of lines and ladder patterns scattered over its walls were old although they didn't know how old. The cave, situated among scattered villages and settlements, some there for thousands of years of human habitation, must have been visited many times before but whoever those visitors were, they felt no need to contact, as the young cave explorers did, the national archaeological authorities who organized its excavation.

In a country rich with the sites of cave paintings, the archaeologists thought them recent, only 2,000 years old, dating from the early Iron Age, new by comparison with the 20,000-year-old depictions in the caves at Nsalu nearby, or the 18,000-year-old paintings at Nachikufu a few miles further north.

The walk we took to the cave was long, through quiet, wild, beautiful places of soft earth paths and stillness, of rustlings and scuttlings and new kinds of birdsong. The day we went, I felt an unfamiliar sense of being in a place not dominated by humans – just being human seemed an intrusion on others' space, on time itself. Inside the cave, the faint ochre scores and lines were markers of the lives of long-ago humans, what they'd seen and what they'd drawn. Time seemed to spin away into clear light until nothing intervened between us and the painters of the cave. There, it was easier to imagine the world as we evolved into it, an already-old world populated by species already ancient, of a time when we began to make images of what we saw, of sight and mind, how we wove the land, sky, stars, other living beings into the explanations we needed to understand the questions we still ponder about where we came from, and how we got here.

The narrative of where humans originated and how we spread across the world is, like our developing knowledge of our origins themselves, a story that is constantly changing. Between 80,000 and 60,000 years ago, there were probably several species of our hominin ancestors spread across Africa and Eurasia, with *Homo sapiens* possibly arriving in Australia 60,000–50,000 years ago, and in the Americas 15,000–13,000 years ago. In many places humans and other species – including the larger species such as mammoths, mastodons, camels, rhinoceroses, cave bears, cave lions, elephants and many others – seem to have lived side by side for thousands of years but in others, following our appearance in their previously human-free environments, large mammals became extinct, particularly on islands where different and unique geographical and environmental circumstances affected species numbers and mobility. Much of this extinction is thought to have taken place 13,000–10,000 years ago after the Last Glacial Maximum when, in a warming climate, humans and animals alike had to adjust to a changing world.

The question of why and how much of the world's Pleistocene

megafauna became extinct preoccupies a lot of people in the same way as the fate of dinosaurs. It seems to reflect and characterize our attitudes to other species and the role humans may have played in their lives and deaths. The larger creatures have always exerted a particular fascination for humans, one that may have been detrimental to their well-being as we've both admired and exploited them beyond the limits of their sustainability. It was in Zambia that I saw one of those designated megafauna in its own habitat. We were wandering near a river at night listening to night calls, trying to identify footprints in the river-mud when an elephant passed us quietly, almost without our noticing, accompanied only by the lightest rustle of grass, a moment of darkness against darkness and softly lifting feet.

In North America between 12,000 and 10,000 years ago, thirty-three genera of megafauna disappeared, including mammoths, mastodons, ground sloths, camels, dire wolves, short-faced bears and other large mammals. In Australia, 15,000 years ago after humans arrived, thirteen species of large megafauna became extinct. Large lizards, giant kangaroos, marsupial tapirs, large echidnas, terrestrial birds and marsupial lions all disappeared. South America's megafauna appear to have overlapped with humans for thousands of years before the extinctions of fifty-two genera 12,000 years ago, including sabre cats, horses, glyptodonts, mastodons and large armadillos.

In 1967, a book written by the palaeontologist Paul S. Martin was published that proposed a hypothesis to explain the phenomenon in North America. *Pleistocene Extinctions: The Search for a Cause* introduced the concept still known widely as 'the overkill hypothesis'. According to Martin's theory, North American megafaunal extinctions happened some 13,000 years ago as a result of the hunters of the Clovis culture carrying out swift and devastating assaults on 'naive' species – those unused to the presence of humans and therefore more vulnerable to attack – using the sharp, fluted stone tools known as 'Clovis points', which are among the very few traces of itself the nomadic culture left behind. The 'points' themselves are beautiful, distinctively flaked, shaped and sharp, forming spear-tips and perfect oval knives. Made from jasper, obsidian, chalcedony, chert, agate, quartz of deep red, mottled black, soft ochre, pearl and smoky grey, the perfection of their making seeming as carefully worked as treasured caches of jewels. The other traces of the Clovis were slight – evidence of a diet that appeared to have relied on small animals, plants, fish and birds' eggs; the 12,600-year-old remains of one infant found in Montana and fourteen 'kill sites' where the bones of mammoths and mastodons were found together with Clovis points.

The major area of dispute surrounding Martin's hypothesis was the lack of archaeological evidence to support it. There are those who believe that fourteen kill sites was a large and significant number but others who disagree and suggest that, were his theory correct, there would be many more. Martin's own assertion that his model neces-sarily involved a lack of evidence, seems to leave unresolved the entire question and many more, including why many other species living at the same time, such as bison, moose, elk, caribou and others, survived.

Plenty of other theories exist to challenge Martin's. One is that the extinctions were caused by a 'hyper-disease' introduced to other species by the newly arrived humans or their accompanying domesti-cated animals, and while no evidence exists to support the theory (and finding any would be challenging) it would explain the relative speed of the extinctions. Another theory is that species who had become adapted to colder temperatures were unable to survive the later climate warming, which encouraged the expansion of human populations and changes in vegetation. Many favour the suggestion of slower processes being involved in the extinctions, from the anthropogenic effects of deforestation and habitat loss, or the long-term repercussions on species numbers of the killing of even small numbers of large animals who have slow rates of reproduction.

There are those too who believe that extinctions may have begun before the arrival of humans – some species have been shown to have begun declining 25,000 years before they became extinct. Others point out that it would be unlikely that Pleistocene hunters would have been unaware of falling numbers of large prey or of the consequences of the depletion of their own food sources. At the time of Martin's writing, there was no evidence for human presence in the New World predating the Clovis people. In the 1970s and 1980s, the discovery and excavation of sites of earlier human habitation significantly altered the view of subsequent events, with some research suggesting that most megafauna, already almost extinct upon the arrival of the Clovis people, underwent two major declines before total obliteration. The same research also suggests that there's no evidence of early hunting in certain areas of north-eastern North America, lending weight to the idea that environmental and climate factors were the more significant in species extinction.

Although Martin's book was written many years ago, his hypothesis remains contentious and its influence continues. In one way, the words Martin used seem the most significant factor, gaining their own momentum over time – 'overkill' and 'Blitzkrieg'. 'Lightning war' is an expression that overlooks the fact that our Pleistocene ancestors weren't the murderous savages who employed this particular military strategy in furtherance of their aims of genocide and conquest. Since the book's publication, much of the debate surrounding the idea has been expressed in the language and atmosphere of the detective fiction or the courtroom drama, 'Is Paleolithic Man guilty as charged?', 'Is this the smoking gun?' Although not Martin's intention, the theory has been used both to excuse and justify aggression and misogyny and bolster the erroneous belief that we're little changed from the way some like to believe our early ancestors were and that 200,000 years or so have had no altering or modifying influence socially, culturally, genetically or in any way that might lead us to desist from behaving badly.

Everywhere humans went, we marked our human presence as we

did in that little cave in Zambia, in lines and circles, in drawings on walls. Among the first things we drew or carved into rock and stone were other creatures, close co-inhabitants of our earlier world. Their images are still there in the caves at Bhimbetka and Daraki-Chattan in Madhya Pradesh where tigers and elephants, wild boar, snakes, lizards and peacocks are depicted among the carved and abraded petroglyphs and cupules scratched into the rock sometime between 700,000 and 290,000 years ago. In caves in Sulawesi, among the 40,000-year-old handprints, there are images of wild pigs and little, horned and hairy deer painted in ochre.

In 1994, three French speleologists, investigating a draught blowing from a cleft in rocks at Chauvet-Pont-d'Arc in the Ardèche, cleared their way into a cave entrance concealed by a rock fall for 25,000 years. What they found there is still considered to be a supreme aesthetic achievement of early art – 400 depictions of lions, horses, reindeer, musk oxen, rhinos, bears, ibex, bison, panther, aurochs and one small, engraved long-eared owl turning her head in the 270° rotation common to fixed-eyed, bendy-necked owls everywhere. Painted over a period from 37,000 to 28,000 years ago, the drawings are all beautiful but the depiction of the horses in the perfection of their movement and expression suggests a deep affection and familiarity between human and horse that seems to connect us in an unbroken line with the consummate artists who drew them. Fragments of the charcoal used in their creation still scatter the cave floor beside them.

No one will ever know for sure why these creatures were drawn or what role their creation played in the lives of their creators. No remnants of habitation were found in the cave but however much we imagine, theorize and speculate, we'll never know why they went there specially to create work that may have been decorative, symbolic, ritualistic, mystical, totemic, shamanistic or an expression of an altered state of consciousness induced by the high carbon dioxide levels in the air. Cave walls themselves have been seen as boundaries between the realms of one sort of consciousness and another but the people who decorated them may have been little different from us in whatever impelled them to draw and to decorate – perhaps simply finding pleasure and fulfilment in observation and the exercise of skill.

As soon as it was found and examined the cave was sealed by the French government to prevent the irreversible damage that has all but destroyed the paintings in the better-known caves at Altamira and Lascaux. In those caves, exposure to humans and their attendant bodily secretions has led to the development of funguses and moulds, which now cover the walls. The high levels of radon and the carbon dioxide inside the Chauvet cave, which may have induced hallucinatory states of mind in the painters, make it unsafe to be in for more than brief periods and only very limited access is given to archaeologists, scientists and artists. Among the latter were John Berger and Werner Herzog, both of whom made films of their visits: Berger's *Dans le silence de la Grotte Chauvet* and Herzog's *Cave of Forgotten Dreams*. In both films, the narrations are quiet, almost reverential in the total silence of that immense and eerie space of glittering calcite, of stalactites and folded rock. Floors are indented where cave bears slept, marked by their claws, padded with paw prints and footprints, including the footprint of a child. On one slab of rock, the skull of a cave bear remains, precisely placed. Walls are scattered with handprints, one in particular recurring, distinctive because it was made by a person with a crooked finger.

Herzog describes the cave as 'the greatest discovery in human culture', seeing the careful siting of the images and masterful use of the surfaces in their interplay with lines and light and shadow as 'proto cinema', beyond the static in the subtlety of their fluid movement.

For Berger, the cave manifests the human relationship with other species. He reflects on how different this relationship was for the people who created this place of beauty, born into a world where animals were supreme and the vastly more numerous 'keepers of the world'. He records the respect and pleasure with which he believed the animals were depicted and considers the skill comparable with the work of Fra Lippo Lippi, Velázquez or Brâncuși: 'Apparently,' he says, 'art did not begin clumsily. The eyes and hands of the first painters were as fine as any that came later. There was grace from the start.'

His words and his belief that the creators of these beautiful images were people of 'grace' touch me. The view is unlike that of early humans as the 'caveman' of cartoon and popular belief, little more than a thick-skulled, wanton destroyer. These Aurignacian artists of the Chauvet Cave saw space, according to Berger, 'as a metaphysical arena of continually intermittent appearances and disappearances'. However they viewed the world, their world, as ours, was one of change in climate, environment and species, as it is in any place, at any time on a dynamic planet.

The first drawing materials were charcoal and ochre, a natural combination of iron oxides and hydrogen, its pigments depending on levels of haematite and manganese, widely used in funerary rites too. On the Southern Cape coast in South Africa in the Blombos Cave, an artist's workshop was found equipped with ochre and bone, charcoal and fragments of quartz stored in abalone shell containers, grinding stones of quartzite and stirrers made from animal bone, shell beads and pieces of engraved and cross-hatched ochre rock, a workshop possibly 100,000 years old. Archaeologists at Twin Rivers near Lusaka, working a few hundred miles south of the cave David and his siblings found, discovered rock tools stained with colour made by ochres of red and brown, yellow, purple and black, which they believe to have been used 300,000 years ago by *Homo heidelbergensis*.

The only surfaces our earlier ancestors left behind, probably the only ones they had, were the walls of caves, places that lay closed for millennia until they were broken into, uncovered, laid open like long-hidden galleries or secret libraries. The lives of these early artists will always be unknown to us, but like us they were creatures who made things using tools of flint and stone, creatures who used fire, embellished themselves with jewellery and clothing, who used weapons and sewed and wove and made music as the first musicians did with flutes of vulture-bone and mammoth ivory 40,000 years ago. They chipped and engraved and drew and fashioned and spread their hands flat against walls to leave their imprints. They may have been making patterns or stencils, or symbols of esoteric spiritual practices we can never understand. Perhaps it was a way of creating beauty or of simply saying, as most of us have, whether by scrawling or carving our initials or obsessively taking photos of ourselves, 'I'm here, now, at this place and time.' Once, it was assumed that cave art was unquestionably made by men but a study in 2013 measured handprints and hands, taking into account the differences in hand size and appearance, which might be explained by place, genetics, origin, time and every other relevant factor, and the authors concluded that the majority of identifiable prints were female.

For later art, there were canvases, linens, silks, calfskins, plaster and stone to work on. Towards the end of the fourteenth century, Cennino d'Andrea Cennini wrote about ochre in his glorious instruction manual on artists' materials, *The Craftsman's Handbook*. He describes an expedition with his father into the hills of Tuscany to collect ochre: 'And upon reaching a little valley, a very wild steep place, scraping the steep with a spade, beheld seams of many kinds of colour: ochre, dark and light sinoper, blue and white . . . Ochre colour is of two sorts, dark and light. Each colour calls for the same method of working up with clear water; and work it thoroughly, for it goes on getting better.'

The writer and Egyptologist Susan Brind Morrow has written extensively about the close connections between language, religion and nature and the power of words. 'The word carries the living thing across millennia,' she writes, a reminder of how words bridge time, the words of observers, inheritors, tellers of tales, the historians and archivists of humankind's memory on earth. There was grace here too, among the handers-on of knowledge, the keepers and bestowers of humankind's long experience who formed it into the narratives we've always told about our origins on earth. Over the long years of our emergence in the world, we've formulated ideas about how we appeared, evolved, how we might have crept slowly from the microbial soup of sea, how we might have dropped from the sky as the fabric of stars raised or fallen, how we might have begun as luminaries, actors or hapless fall-guys in the machinations of others.

The themes of creation stories are the universals of all life, now as then – darkness, stars, moon, sun, eclipses and auroras and most of all water, the predominant element of the lacustrine and oceanic cultures from which so many of them originated, the flood plains of the Nile, the marshlands of the tributaries of the Tigris and Euphrates, the water-bound cultures of the Pacific north-west; water with its bounties and its terrors, its promise of abundance or death. The first-known written creation myth, the Babylonian 'Enuma Elish', took the form of 1,091 lines of Akkadian cuneiform inscribed on seven tablets, discovered in the ruins of the library of Ashurbanipal in Nineveh in 1849. The story is of the gods before the creation of heaven and earth. It bears similarities and correspondences to the foundational myths of many cultures, including the one that has provided some of the most significant underpinning beliefs of Western thought: the book of Genesis with its resounding opening words 'B'reshit', 'In the beginning . . .' the biblical story of the creation of life, containing within it the echoes of the creation myths of the Sumerians and Akkadians, the Assyrians, Chaldeans and Babylonians. Their stories are written into the tangled epics and rolling narratives of the histories of the region, from Egypt, Greece and lands of the Levant, accounts of floods, of long chains of patriarchy, family history and stories of the origin of all things.

To use the word 'myth' of the creation stories carries a suggestion of untruth but the stories of human origins are untrue only if they're viewed through the unyieldingly rationalist view that rejects the symbolic and metaphoric. On the function of myth, the anthropologist Bronisław Malinowski wrote in 1926: 'Myth is thus a vital ingredient of human civilization: it is not an idle tale but a hard-worked active force; it is not an intellectual explanation or an artistic imagery, but a pragmatic charter of primitive faith and moral wisdom.' Although the word 'primitive' is no longer used of past life or belief, the continuing dynamic significance of the cosmogonic – creation – story is powerful. In many of these stories, other species are represented as neither antagonist nor prey but as gods, creators, or as intermediaries between the divine and human worlds, or as mortal figures who possess immortal powers. They may be instigators, shape-shifters, innovators, creators who play easily with our dull-witted species.

In Richard K. Nelson's *Make Prayers to the Raven*, an ethnographic study of the lives of the Koyukon of the northern forests of Alaska, he writes of Distant Time, the beginning of time itself, and of the origin stories which encompass earth and sky, topography and weather. Wind has a spirit that can be pacified by certain actions (such as not behaving disrespectfully towards grey jays). Thunder, as the embodiment of a human spirit from Distant Time, has awareness and consciousness. Each creature too, each bird and insect, has spirit and significance emanating from Distant Time. The accounts and stories are peerless, often witty and sly, loud with the overlapping whispers of ancient memory. There are many types of creation myth – the near-ubiquitous flood story, or the 'fall-of-the-sky' story, the 'out of chaos' and 'earth-diver' story, all telling of differing ways of the world's beginning. There are those that interweave the lives of humans, animals and birds with a cast of characters as beguiling in name and activity as the most ebullient superhero. In the creation accounts of the Tsimshian people in the Pacific north-west we encounter 'The One Who Walks All Over the Sky' and his brother 'Walking About Early'. Both appear together with their sister 'Support of Sun'. There's brilliant, resourceful Anansi of the Akan, a figure of moral import who has expanded beyond his origins in Ghana to become emblematic throughout Africa and the Caribbean in the form of a spider, a trickster, son of the Sky God Nyame and Asase Yaa, the Earth Goddess. Coyote, an almost ubiquitous figure in Native American accounts, is both hero and anti-hero, the embodiment of opposites, above all a survivor. These are qualities shared by the greatest creation myth character of all, Raven.

A few years ago, the American Public Broadcasting Service showed a short film as part of a series on corvid intelligence. Filmed in Scandinavia in the deep snow of winter, it documents the bewilderment of a fisherman whose ice-fishing catch is stolen repeatedly. As he sets his fishing line over the hole he's cut in the ice, he's being closely watched by a nearby raven who, as soon as he leaves, flies down to retrieve his line, pulling it up with her beak, securing it with a foot until she lands the fish. The moment when the fisherman returns to find the raven in possession of his fish is salutary and very funny as he rages ineffectually at the bird's swift lift-off with a large trout in her beak. In this short scene is every explanation of why ravens form such a large part in the creation narratives of many cultures, appearing in every mythology, or every one originating in a place where corvids are common, Scandinavia, the Baltics and Hungary, the territories of Celtic tradition although most stories come from the cultures of the Pacific north-west and Alaska, from the Haida and Tlingit, Kwakwaka'wakw, Koyukon, Salishan, Nisga'a and Tsimshian peoples – 'How Raven Steals the Light', 'How Raven Frees the Light', 'How Raven Gets Caught in a Lie', 'How Raven Invents Fire', 'How Raven Loses his Beak', 'How Raven Makes the World'. In *Tulugaq: An Oral History of Ravens*, Kerry McCluskey's interviews with people across the Canadian Arctic reveal account after account of raven legends passed on by grandmothers and elders, as well as daily observations and personal encounters with the birds. It's clear that these stories have long constituted a basis for the teaching of social laws and moral behaviour.

As our relationship with other species evolved over time, the way we represented them did too. Our depictions of animals and birds expanded beyond the confines of the cave and its enclosed and secret walls. As our societies grew everywhere, the way animals were portrayed altered in form, in expression and in the materials used as developing cultures employed media beyond the limited materials of early art – ochre, charcoal, shell, ivory, horn and bone which, with the development of metallurgy, allowed the use of bronze and iron, gold and silver. Ceramic cultures developed in China as early as 20,000 years ago and 12,000 years ago in Japan and in parts of Europe where vases, drinking vessels and statuettes began to be decorated with zoomorphic representations of cows, goats, deer and hedgehogs.

Stones too were worked, sculpted, engraved, often depicting the lives of humans and animals together. Around nine thousand years ago in the Sahara at Dabous, in what is now Niger, two large giraffe petroglyphs were carved, the larger 6.35 metres long, as well as other animals, lions, ostrich, antelope and rhino. In China 6,000 years ago jade carving began with carefully polished and incised tools and axe heads; the culture of Liangzhu produced finely crafted figures of birds, fish and dragons. In the settlement of Catal Hüyük, in modern-day Turkey, dating from 7500 BCE, a place now regarded as a significant transitional stage between the nomadic and the settled, paintings and reliefs of animals and birds – among them bulls, vultures and leopards and some incorporating animal teeth and horns – were found together with remarkable hunting scenes which seemed to portray and suggest complex relationships between human and animal, both domesticated and wild.

Between 5000 and 1500 BCE, in the decoration and art of Mesopotamia, at Tepe Gawra in present-day Iraq, goats, snakes, cows, calves and deer were among the creatures depicted on the seals which have been described as 'the interface between writing and art'. In Nimrud, winged, human-headed bulls or winged lions appeared in hunting scenes with their huge, majestic depictions of animals. In the palace of Ashurnasirpal II, beautiful carved gypsum panels depicted horses, oryx, lions and, memorably, a winged man holding a deer and a branch. They're there in the form of the bulls and dragons of the Ishtar Gate of Babylon, the lapis and silver bull's head from the royal tombs of Ur. In the capital of the Hittite Empire, Hattusha, huge stone lions guard the gates, seeming to smile with sweetly benign and toothless smiles, as animals became symbols of power, their qualities mirroring and reflecting back the power humans took upon themselves as rulers, masters, emperors and kings. On the walls of the Apadana, the hypostyle hall, audience chamber of Darius and Xerxes, in Persepolis, capital of the Persian Empire, tribute processions lead the animals of their realms, fine Scythian horses, Bactrian camels, sheep, donkeys, goats in a never-ending line.

Memory and voice carried the words until writing developed in Sumeria around 3000 BCE. In a stern political and social system that controlled the lives of the Sumerian peasantry, the cuneiform system of writing was the fortunate by-product of the Sumerian accountant's art, expanded from the system of counters used to tally goods. After millennia of ochre and lines, shadowy handprints or faceted points of stone, writing moved humanity into a new phase with our recording of our deeds and thoughts, marking our lasting, individual presence on earth.

Alongside the monumental artwork of Sumeria, Babylon and Assyria, animals appeared on more domestic and personal items – drinking vessels, ornaments, jewellery, the sash pendants and finials, belt hooks and bracelets. Cylinder seals too, those part-amulet, part-jewellery, part-identity cards of ancient society widespread throughout the region of the Fertile Crescent, depicted gazelles, lions, snakes and other creatures, real and fabulous, made from lapis lazuli, faience, carnelian or amethyst. It is the apparent sympathy between human maker and beast in the small artefacts that beguiles, when there's no grand political gesture to be made, no self-aggrandizing, no co-option of the symbolic powers of other species. The respect and humour with which the makers of the objects drew inspiration from the natural world is apparent in the details: the small, 6,000-year-old, boggle-eyed Egyptian pre-dynastic elephant amulet, the 5,000-year-old Yangshao eagles with the anxious eyes, the 2,500-year-old Palaeo-Eskimo Dorset culture carved ivory fish, polar bears and seals.

I reach out to pick up the small Victorian clockwork finch of soft brown plush who stands on the bottle of ink on my desk. He's just one of the representations of other creatures which surround me, no different from those ancient artefacts in the reasons I appreciate them: the celadon cow with the broken horn, the cot-toy dove, the finch, the Christmas-tree decoration white fox with his red felt scarf, symbols of memory, appreciation, love.

The complex boundaries between human and animal worlds, temporal and spiritual of Egypt are woven into the depictions of divine power shown through ibis or baboon, cat or crocodile, hawk or goose. The exquisite scenes of the life of the Nile from the tomb-chapel of Nebamun in Thebes, now in the British Museum, show creatures in such detail that they fly and stalk and swim, the marmalade cat lurking in the reeds with the bird in his claws, the geese collected into a basket, the wagtails, hoopoes and quails in flight, the tiger butter-flies, the hares, fish, gazelles, all so perfect and abundant that they're suffused by the illusion of sound, the whirring of wings and the calls and cries of birdsong, the lap of water in reed beds, the heavy hum of insects in scented air. Dreamscape and vision, the frescoes portray life and afterlife, the representation of impossible earthly perfection in the garden of a wealthy Egyptian in this life and the next, resonant with associations of gardens and paradise – the word derived from the *para-deisos, paradaijah, pardes* of Greek, Persian and Hebrew – the sacred in life on earth or beyond, in a garden of Eden, a garden of animals, a divine and heavenly orchard.

The delicate depiction of an Aurignacian owl or the butterflies and geese of Nebamun's chapel, like the grandiose conceits of Mesopotamia and Rome, are part of culture, society, a natural world observed. They're expressions too of mystical possibility, admiration, of zoolatrous worship or totemism – the belief in kinship with a sacred object – or pointers to hybrid worlds and connections between humans and other species. The still-mysterious bull-leaping frescoes and arte-facts of the Minoans may be imaginative portrayals of cosmology or athletic rites of passage but no one can be sure. The strange and beautiful anthropomorphic animal figurines of the Proto-Elamites, 5,000-year-old incorporations of aspects of human and animal in the half-smile of a reclining silver goat, the solemn-faced, kneeling bull offering his beaker, tell of affection, intimacy, recognition. In her many portrayals, Britomartis, the Minoan figure of 'Queen of the Wild Bees', or 'Mistress of Wild Animals', called too by her Mycenaean name 'Potnia Theron', appears surrounded by animals, her ubiquitous griffins, lions, bees, snakes and birds, expressions of the connections, actual and symbolic, between species, another thread through time and the territory of our imaginations, loves and fears.

We've always been entwined in life and in death with other creatures although often,, too much time has elapsed to be able to interpret with any certainty what some of the symbols and artefacts mean. They still lie in the darkness of caves and graves, our older selves living on in what we drew and what we left, in how we died and how we were buried, in the testimony of our secrets, enmities, cruelties and terrible griefs, the kind which reveal the similarities and differences of our all too human state. They're still there in our ornaments and grave goods, often the remains and parts of other creatures – pendants made from bird bones, necklaces of shells and teeth, the bones of sheep, rabbits, fish, the wing bone of a golden eagle, the skeleton of a white-tailed eagle, the leg bones of a goose, used as decoration or amulet, possessions, totemic symbols or offerings to the unknowns of a possible future life.

The archaeologists who discovered a fox buried with a human in a 16,000-year-old burial at Uyun al-Hammam in northern Jordan, the oldest cemetery ever found, believe the finding suggests that there was a relationship between animal and human – the fox was buried complete, daubed with the red ochre associated with human burials, indicating the possibility of its having been a companion in life and an intended one in death.

In the Caverna delle Arene Candide on the Ligurian coast, the wing of a corncrake was found laid on the chest of a child; on the chest of another, two beaks and the single wing of an Alpine chough. Among the layered burials dating from the Byzantine Empire (330–1453 CE) back to the Gravettian (an Upper Palaeolithic European culture which may have extended from 32,000 years ago until 20,000 years ago) numerous animal remains were uncovered – the left back paw of a wolverine, the leg bone of a beaver, the skeleton of a red deer, and parts of goosander, herring gull and red-crested pochard. Below later layers, the 23,500-year-old skeleton of a Gravettian boy estimated to have been about fifteen lay on a bed of red ochre, richly adorned, a cap made from hundreds of perforated shells and the teeth of deer around his head, his body decorated with pendants of mammoth ivory, pierced elk antlers and in his right hand, a long blade of flint. (He's been called 'Il Principe' because of the richness of his adornments and the care with which he was buried. With a neck wound concealed under layers of yellow ochre, he was already suffering from advanced Pott's disease – spinal tuberculosis – which had destroyed several of his vertebrae, one of the earliest cases of the disease to have been found in Europe.)

In a Natufian burial at Hayonim Cave in the Upper Galilee, dating from 13,000 years ago, three humans were interred with two small dogs while nearby at 'Ain Mallaha near Lake Huleh, in another similar site, a woman was buried with her hand placed on the puppy beside her.

Over two thousand miles to the north at Vedbaek in Denmark, a very young woman was buried, similarly richly decorated. She was garlanded with necklaces made from animal teeth and snail shells around her head. Her face and pelvis were scattered with red ochre, and beside her, on the wingtip of a whooper swan – specifically on its right carpometacarpus – lay the body of her newborn son. Like other waterbirds, swans play a large part in the mythology of the northern countries. All migratory birds carry messages, their departure in autumn and return in spring possibly heralding more than just a new season with its promises of vitality or stillness. The symbolism of flight and transcendence gives an unforgettable poignancy to this possibly 7,000-year-old Mesolithic Ertebølle burial, the too-early deaths of mother and child. Nearby graves contain other animal parts: a grebe's beak, a pine marten's foot, a dolphin vertebra, a set of deer antlers. Whatever the meanings of the offerings, the fact that they're there at all opens a small window onto the immensity of grief and the place of other species in its expression.

The Pleistocene dog buried with two humans at Bonn-Oberkassel 14,000 years ago might have been just another dog. Originally found a hundred years ago and recently re-examined, he was discovered to have been young, probably around seven months old, and to have died from distemper. Examination of his teeth suggested he'd been suffering from the almost invariably fatal canine disease for some weeks and had probably survived for as long as he did because he was being carefully looked after. The conclusion of the archaeologists who examined him was that he must have been a pet, precious in some way to whoever nurtured him thousands of years ago. His presence in the grave tells us not only of love and care towards a member of another species but also of one of the most significant processes to take place between humans and other species, that of domestication.

Domestication may have begun long before the Neolithic Revolution but it gained momentum at that crucially important time in human history. With the post-glacial climate warming at the beginning of our epoch, the Holocene, agriculture expanded worldwide and the relationship between humans and other species began to change. The ways by which domestication might have begun have been suggested by the archaeologist Melinda Zeder, as being by commensalism when wild animals began to take advantage of the food resources of human settlement, animals such as camels or reindeer being captured and bred or when those previously hunted were kept to be domesticated. The processes of domestication are prolonged – selective breeding modifies and alters an animal's characteristics by bringing about behavioural, genetic and morphological changes, which allow them to be used for the benefit of humans. Much of the process began in western Asia in the Fertile Crescent, the area of the Nile, Tigris and Euphrates rivers and the land now constituting the countries of the Middle East, parts of Turkey and Iran, then in India, China, the Americas and worldwide.

Wolves were probably the first animals to be domesticated (possibly as early as 40,000 years ago), to be amenable to our will. Sheep may have been domesticated 11,000 years ago, and were followed by cats, goats, pigs, cattle, guinea pigs, horses, camels, zebu, water buffalo, chickens, ducks, geese, turkeys, reindeer, yak, doves and many other species. The relationship between humans and other species changed for ever as, one by one, these creatures were domesticated. No longer simply co-habitants of the earth, they became a 'managed resource', exploitable and dependent as humans established ownership and with it total control of every aspect of their lives.

A few weeks ago, it thundered. I woke to it, the bedroom window open after a hot day. At first, the sounds were low and deeply resonant, expanding slowly into magnificent, crashing, Wagnerian, voice-of-the-Almighty, Old Testament loud and fully orchestrated thunder. There always seems to be a terrible warning in it. This is the voice, yet another manifestation of the natural world. Take me seriously. I lay and listened, thrilled and impressed by the power of this phenomenon of electricity. I got up to close the window before the rain began. An opaque black sky flashed brilliant-blue and luminescent, overshot by dazzles of pure white light. I heard Chicken wake and call from downstairs. 'It's OK,' I called back, 'it's only thunder,' and I heard her chair move slightly as she settled back to sleep. I went back to bed and waited for the rain, which began as powerfully, as majestically as the thunder. It crashed, bounced off the pavements, tarmac, car roofs, flowed in the gutters from a different place, from 'the country of the thunder people' of the stories of the Oneida people, from the will of those thunder gods, Zeus and Set and Thor, from the maker of a different world, Raven, creator of sun, stars, moon, fire and water, creator of the world and bringer of light. I lay awake, wondering. Did the world begin this way? Did we waken one morning into a watery world newly created by a raven with a sense of humour? I slept, and wakened to a cool day, grey and uncertain, to low rumblings from a heavy, clouded sky.

The American anthropologist and archaeologist, philosopher of science and writer of profound and lyrical insight, Loren Eiseley, a habitual insomniac, did much of his work during hours of wakefulness. In his book *The Immense Journey* there's a remarkable account of his waking in a room on the twentieth floor of a hotel in New York before dawn and watching from the window as the city's pigeons rose from their high lofts and roosting places, reflecting light from their wings as silently they flew out over the city 'on their mysterious errands'. His sense of being part of that other world is clear when he expresses his almost overwhelming wish to join the birds by launching himself as a young bird might, out into 'the city of wings'. Far from being an account of despair or the wish for self-annihilation, it is one of identification and his ever-present awareness of the mutable boundaries of otherness. His calling of the pigeons 'these other inhabitants', shows that he regarded all the other creatures he encountered as extensions, fellows, equals.

In 1935, while a student at the University of Nebraska, Eiseley took part in the excavation of an Ice Age site at the Lindenmeier Ranch in Larimer County, Colorado. The dig was conducted under the aegis of the Smithsonian Institute and Dr Aleš Hrdlička, its stern founder and curator of physical anthropology. Hrdlička believed that humans had been in North America for only 3,000–4,000 years, having crossed by the Bering land bridge, the now submerged land corridor thought to have once joined Asia and North America. This was the 'Bering Straits theory'. Hrdlička's conviction was more than it seems. A Czech emigrant to the United States, he spent most of his academic life trying to prove the superiority of 'the white races', subverting science to the ends of 'scientific racism' by skull-measuring, limb-measuring and spurious data collection on physical and supposed moral and intellectual differences between races. It was couched in the usual hateful terminology: 'abnormal', 'degenerate', 'debased', 'immorality'. Deeply colonialist in intent, Hrdlička's aim was to find 'the American type'. He argued for the immigration most likely to correspond to it by undermining any claims of long habitation by the original peoples of the continent. A prolific collector and storer of other people's bones, Hrdlička died in 1943 before the full and lasting implications of his work could be known.

During the dig, it was Eiseley who made the most significant find, that of the vertebra of a long-horned bison, *Bison antiquus*, with a Folsom point (one similar to, but from a slightly later culture than the Clovis) embedded in it. It demonstrated that there had been a human presence in North America since the Ice Age, proving Hrdlička comprehensively wrong. The excavations continued until 1940, unearthing 51,000 artefacts, among them beads made from bone, haematite and coal, drills and awls, knives and fine bone needles, and continued until the economic demands of America's participation in the Second World War cut the science budget, bringing a permanent end to the project.

Forty years later, Eiseley, by then a distinguished academic and poet, wrote an account of travelling back to Colorado by plane through a blizzard in his poem 'Flight 857'. Always deeply conscious of the contradictions implicit in working across the divide between art and science, Eiseley managed to operate in both worlds with unusual ability and compassion. Recalling what they'd found at Lindenmeier and what they hadn't, he expressed an unease that accompanied him all his life about the endeavour of laying bare the past, wondering what had been achieved by this revealing of secrets. It seems as if looking down into the blizzard had made him see, as through a 'lost doorway of snow' the briefness of his own life and the length of the silence into which time would enfold him too.

One afternoon last year when it was nearly winter I was walking near a small loch with a friend. We watched as a faint frost coated the fields and trees and knew that it was the first sign of the real cold beginning. A low and lambent sun cast everything in powdery gold – lines of late autumn geese overhead were underlit in gold, a pair of swans glided on a pond of gilded rose-pink water. It was easy that day to understand the ancient sense of interconnection between the natural and supernatural, human, bird and animal together in waiting equipoise, between worlds. What did the swan's wing placed under the dead infant mean? Did it represent the carrying of the child to the afterlife on a wing, symbolic of the connection between the worlds of life and death? Was there special significance in the wing of that particular bird? Did the whooper swan represent the soul? Was it a gesture, undertaken from the helpless, inchoate desire to provide comfort, even in death, to a tiny baby? The image of the child on the wing is beautiful and seems like a bridge between worlds, times, beliefs. I thought of Eiseley's words and what we can never know of what it was we found, of what might have been the beliefs and hopes of the people of Vedbaek and Arene Candide, reminders perhaps of our limitations, that in our own otherness, we don't know the mind of our neighbour, the true intention of our friend, the disposition of our lover and that in the ways we use and portray other creatures we can only reflect back to ourselves who we really are.

3

Souls

It's a cool spring day when I pass a farm bonfire, just some fallen branches stacked and smouldering. No one seems to be tending it but it's not going to spread – the ground is too damp and the flame too reluctant. Smoke lifts, spreads across the hill above, dissipating into the brisk spring wind. As I'm passing, it hazes into an image from a film I've just seen, the strangely beautiful *Le Quattro Volte*, which opens to smoke blowing through the tops of trees and across the hills of Calabria, rising from a perfect construction of wood, a wide, high dome, built according to the charcoal burner's ancient art. According to its director Michelangelo Frammartino, *Le Quattro Volte*, 'The Four Times', was inspired by the landscape of Calabria where he was born, and by the beliefs of that most enigmatic of Greek philosophers Pythagoras about the stages of the transmigration of souls. These ideas, attributed to Pythagoras, himself a one-time resident of Calabria, infuse the film's portrayal of life, birth and death and the apparently insubstantial boundaries between human and animal, vegetable and mineral.

The film is an account of the life of an old man, his herd of goats and his dog, set against the felling of a magnificent tree destined to be made into charcoal. Its background is of natural sounds, of goats, birdsong, insects and – only faintly – of humans. The faces of the goats with their look of calm nobility are given the same time and weight as that of the goatherd, the only human to play a substantial role in the film. Goats are the stars. Ants and flies too, the snails escaping from the old man's cooking pot, his brisk, endearing dog (who, for his performance, won a special jury prize, the Palme Dog, at the Cannes Film Festival). Throughout the film, it's almost as if dust and smoke obscure the sharp lines to which we're accustomed, effacing the boundaries between species, in part of a process that has no end. A small white goat wanders perilously on her own and an old man dies. The life and death of human and non-human alike are mirrored in the transformation of tree to fuel, part of the mysterious, immutable journeying of substance and of souls.

It's difficult even to imagine how anyone first thought of such oddly elusive things as souls, so romantic in their immateriality, so way-wardly defiant of definition, no more than airy, ethereal constructs, mystical yet endowed with indestructible power, both passionate in-dwellers and enduring, long-distance travellers at the same time. Yearning representatives of our all too human longing to be more than our worryingly temporary and fragile selves, souls may express our hapless desire for something, however intangible or unlikely, to continue after our deaths. Our fears and hopes are concentrated into one diversely interpreted idea, an unseen, unassailable, permanent, enduring force which, according to some, just *is*, prone to gathering names, creating fusions, bifurcating, growing, expanding into myriad strands of thought and possibility.

The origins of our phantasmic, liminal companions lie in the ancient Egyptian preoccupation with the afterlife and their compli-cated soul formations. At some times in early Egyptian history, it was believed that the soul had five or seven parts, but the most common belief was that it had nine, connecting body, name, soul and essence in differing relationships: *Khat, Ka, Ba, Shuyet, Akh, Sahu, Sechem, Jb* and *Ren*. In Greece, ideas of the soul, its nature and composition, developed over time, from the ephemeral *psyche*, to the later ideas of Aristotle, expressed in his great work *De Anima*, 'On the Soul'. In Sufism, the soul has seven aspects, representative of body, mind and spirit in relationship to heaven and earth. The Islamic *nafs*, 'soul', is composed of three elements, *nafs-ul-ammarah, nafs-ul-lawwamah* and *nafs-ul-mutmainah*.

In many cultures, the intangible nature of souls is represented by words of air and wind, Hebrew's *nefesh, ruach, neshama* – 'soul', 'wind' or 'spirit', and 'breath', and a further two, '*chaya*' and '*yechida*', for those of a mystic bent. There's '*pneuma*' too, the combination of air and fire of the Stoic Chrysippus. Being aeolian, windy things, souls seem much given to travel, hitching rides on passing birds or under-taking post-mortem, subterranean travel to underworlds by way of rivers Acheron, Styx or Hubur where they gather like insubstantial package-holiday tourists, waiting for crossings to Hades or Inferno, Gehenum or Tuonela, places often reached only with the kindly, or perhaps not so kindly, intervention of boatmen or ferrymen, Charon, Hamar Tabal or Manannán Mac Lir, those watery servants of Death.

From the right carpometacarpus of the whooper swan of the Ertebølle Culture to the soul-bearing cranes of Chinese poetry, other species have occupied a special place in relation to our beliefs about souls, an enduring significance expressed throughout history in early religion, mythology, cultural practice, in ritual and burial rite, in art and literature. The human soul is often associated with other species, in particular with birds. This symbolic connection has early origins – the stork is a hieroglyphic symbol for the soul, and doves too have a long association with the Mesopotamian goddess of war and sexual love Ishtar, with the Greek goddess Aphrodite and with the story of Noah. Being endowed with the enviable ability to fly, birds have been recruited into the service of mystic transportation and message-bearing. Not only the majestic members of the genus *Cygnus* but also the night flyers, the nocturnal, the silent, the stealthy – owls, long distance migrators, those possessed of black feathers and the utterers of notably eerie cries such as the great northern diver, *Gavia immer*. Many seabirds are associated with spirits and souls as intermediaries of one sort or another between the worlds of the living and the dead. Birds are said to appear mysteriously in houses after a death, congregate in unexplained groups around their own dead, carry out their own careful mourning rituals, entering our spiritual worlds as portents, guides or bird-embodied souls.

The motif of the bird as the soul is often employed as a metaphor for escape and freedom in the religious imagery of the Middle East as the soul, unleashed from the confining strictures of the body, is seen as flying free. 'Our soul is escaped as a bird out of the snare of hunters,' Psalm 124 declares. In the poetry of both the eleventh-century Andalusian Jewish poet Judah Halevi and the twelfth-century Sufi mystic poet Rumi, the soul appears to assume the form of the newly liberated bird soaring towards the celestial heights of spiritual destiny in oneness with the munificence of God.

Some believe the soul to be the one element of ourselves to continue to exist after death while others believe in the entire resurrection of the physical being, including those who, in a strangely detailed eschatological quirk, believe that the dead will be resurrected wearing clothes. A seventeenth-century painting by Pietro da Cortona in the gorgeous Palazzo Colonna in Rome, 'The Resurrection of Christ and Some Members of the Colonna Family at the End of Time', depicts the resurrectees emerging tentatively from their graves, the men naked, the women modestly garbed and being helped out by angels. There were those too who adhered to a strand of esoteric Judaism, which claimed that on the day of the Messiah's return special tunnels would appear for amputees to be reconnected with their lost limbs, an idea which brings alarming visions of a Heathrow-style luggage distribution system.

Given the evanescent qualities of the soul, it seems perverse that the possession of one should become one of the apparently inarguable means of differentiating between one species or another, of determining who is worthy, redeemable, infinite, who will be 'saved' and thus will live on beyond the life we know, or not. Throughout history, ideas about who possesses a soul and who does not, have constituted part of the bedrock of the way we've thought about and treated other species. Of the many differences, perceptible and otherwise, between ourselves and other species, it is the idea that has had more influence over our behaviour towards the rest of the living world, enduring through the long philosophical history of disputes and considerations on the ethics of eating animals or the relative cognitive and emotional abilities of other species and consequent moral and ethical responsibility towards them. Much of the significant thought of Western theology makes it clear that it's hugely advantageous to be human and have the variety of soul to be found among humans and generally pitiable to be, and therefore to possess anything else, if only because this lamentable deficiency provides useful justification for any form of behaviour by the former against the latter.

I don't spend a lot of time contemplating souls except when encouraged to by films like *Le Quattro Volte* or when, one morning not long ago, the face of a handsome fox appeared on the screen of my phone, a regular visitor to the garden of the woman who posted his splendid photo and who has named him 'Soul'. By coincidence, later that day, I happened on a website devoted to the doctrinal aspects of one particular branch of Christianity that involved a man answering questions about the fate of the souls of animals. Although some other species may give the appearance of being clever, he said, they aren't really since they lack the capacity to be so. They have no notion of justice, no moral sense and don't live on after death, as humans do. 'There is no doggie heaven,' he wrote, 'unlike us, animals are without immortal souls.'

Others haven't been so certain. Questions and doubts pervaded the writings and accounts of the thinking of the ancient philosophers Alcmaeon, Empedocles, Plato, Seneca, Aristotle, Theophrastus, Plutarch, Porphyry and probably many more before them. These questions were profound and important – did animals have souls? If they did, were they the same as ours? Were their souls immortal? Which qualities and capacities of mind and spirit did animals possess? Were they contained within the concept of 'soul'? Did they have memory? Did they experience emotion? Were they able to think? Were they conscious? Did they have reason? Language? If they did, or perhaps if they didn't, how should that affect human behaviour towards them?

Apart from inspiring *Le Quattro Volte* and being known for his pre-eminence as a mathematician, Pythagoras is often cited as the first to raise his voice in consideration of the nature and lives of animals. In the absence of written evidence, he's been viewed variously as scholar, sage, sinister cult leader, wielder of power or all of them together. His beliefs incorporated what appear now to have been a strange and unlikely set of ideas extending from the transmigration of souls to a prohibition on eating beans by way of other arcana such as not urinating towards the sun, not killing lice in the temple and refraining from walking past a reclining ass.

Bertrand Russell reported caustically, 'It is said that Pythagoras, like St Francis, preached to animals,' which might have been the case, in recognition not only of his reputed habit of vegetarianism but also of his apparent refusal to wear clothing or shoes made from the skins of animals and his choosing not to be anywhere near butchers or hunters. His name seems to be associated with concern for animals because of his belief in metempsychosis – the transmigration of souls. He's said to have stopped someone from beating a puppy because he believed that he heard the soul of a friend in the dog's cries. The third-century biographer of the philosophers, Diogenes Laertius, suggested that Pythagoras believed the soul of man to be composed of three parts, intelligence, reason and passion, while animal souls possessed only intelligence and passion. This question of the possession of 'reason' would become a major one in many future considerations of human–animal interactions and decisions on the appropriate way to treat other species.

Pythagoras's devoted follower Empedocles, as well as formulating ideas on the origin of species, is considered to be the first to envision all matter as being composed of the four elements, air, fire, earth and water. In his poem 'Purifications' he berates himself for the crime of eating animal flesh for which he felt himself to have been punished by the prolongation of the processes of transmigration. For both Empedocles and Pythagoras, the practice of not eating animals may have been motivated at least in part by the more than faintly unnerving consideration implicit in the idea of the transmigration of souls – that in eating an animal, you might inadvertently be eating one of your own relatives.

Although lauded now as a stout exponent of vegetarianism, there's uncertainty about Socrates's view of animals and indeed about his diet. In *The Republic*, Plato suggested that Socrates believed that the vegetarian city would be the ideal, although it's far from certain either that Socrates believed it or that it was on the grounds of the benefits for animals. One suggestion is that, given that the diet of the time was mainly vegetarian, meat-eating was associated with unnecessary luxury and contributed to disputes over land. The later philosopher Porphyry implied that Socrates wasn't vegetarian although familiarity with someone else's dietary habits many hundreds of years later, in the absence of textual evidence, might be subject to question. Plato's thoughts about animals are more difficult to interpret. He employed expressions that suggest that at least he, if not Socrates, saw animals as lesser and lacking in reason, although that appears to be contradicted by his later writings, which talk of animals having both minds and 'the finest sensations' and in the context of his conception of a triadic soul, in which he asserted that the non-rational part of the soul was capable of belief, he appeared to be granting this capacity to animals.

Another of Plato's former students, Theophrastus, proposed a natural kinship between humans and animals based on common bodily make-up and a similarity of soul in impulses such as appetites, perceptions, anger and reasoning, including them within the concept of *oikeiosis*, a word variously translated as 'affinity', 'affiliation' or 'kinship'. He opposed animal sacrifice on the grounds that depriving an animal of its soul was unjust.

Of all the early philosophers, it is Aristotle whose thinking has had the most significant and lasting influence over ideas about humans and other species. Aristotle was different from other philosophers of his day in the scope and depths of his interests, which extended far beyond the bounds of pure philosophy. In his endeavours as a scientist, naturalist and biologist, Aristotle used his close observations of other species in conjunction with his philosophical deliberations to formulate his views of living organisms and of their activities and roles in the natural world. In a lecture delivered at Oxford University on 14 February 1913, entitled 'On Aristotle as a Biologist', the distinguished professor of natural history at St Andrews University D'Arcy Wentworth Thompson described Aristotle with detailed and admiring grace:

> He was, and is, a very great naturalist. When he treats of Natural History, his language is our language, and his methods and his problems are wellnigh identical with our own. . . . He recognised the great problems of biology that are still ours today, problems of heredity, of sex, of nutrition and growth, of adaptation, of the struggle for existence, of the orderly sequence of Nature's plan . . . Evermore his world is in movement. The seed is growing, the heart beating, the frame breathing. The ways and habits of living beings must be known: how they work and play, love and hate; whether they dwell solitary, or in more organised companies and societies . . .

But whatever else Aristotle is, he is the great Vitalist, the student of
the Body with the Life thereof, the historian of the Soul . . .

Aristotle's theories on the soul are intricately complex and his
concept of 'soul' is different from modern perceptions, influenced as
they are by a historic theology that teaches of both immateriality and
immortality. Aristotle viewed the soul as being hylomorphic, a unity of
matter and form, of three kinds, the plant, the animal and the human,
each being the cause of the body's ability to live, each imbued with
differing powers and senses and perceptions.

In *De Anima*, Aristotle proposed that the plant soul is purely nutritive in having 'an internal potentiality' while the animal's is both
nutritive and perceptive: 'all animals can perceive their environment, at
least by the sense of touch. If they have a perceptive part, they also have
desire.' The human soul is nutritive, perceptive *and* rational: humans
have an intellect and the ability to think. While his conclusions seem
to differ from text to text, Aristotle was unwavering in his assertion
that while animals may resemble humans in some ways, they will
never be equal to them because of their lack of reason and meaningful speech. Only humans, he believed, have the ability to experience
mutual affection and through language the means of communicating
values. Aristotle further suggested that since nature is purposeful,
animals must undoubtedly have been made for the use of man, and
man is free from any obligation towards them.

The repercussions of this belief have been profound, as has the
influence of his classification of species, what came to be known as
the Scala Naturae, or the Ladder of Nature, the systematization that
designated not only the positions of species in relation to one another
but the nature of the relationship. According to this system, all of us,
from the 'Blooded vertebrates – Viviparous quadrupeds', that is the
land mammals on the airy topmost rungs, further down into the lower,
darker reaches where 'Birds, Oviparous quadrupeds, Fish, Cetaceans'

belong, and still further down to where the 'Bloodless' invertebrates, the 'Land arthropods, Aquatic arthropods, Shelled animals, Soft animals, Plant-animals' are placed, all the way to the foot of this unforgiving and hierarchic scale where the lower plants and inorganic matter dwell.

Aristotelian ideas of human superiority were cemented into future thinking by two schools of thought: Stoicism, founded around 308 BCE by the Cypriot Zeno of Citium and expanded on by his followers, and Epicureanism, based on the teachings of Epicurus. Both schools incorporated Aristotelian ideas into their formulations for how humans should achieve the state of *eudaimonia* – the beneficial or flourishing life. The Stoics considered that this could be achieved by the practice of *apatheia*, living in a way that freed oneself from the strictures of 'passions' – in their terms, a harmful set of strong emotions, by rationality, inner control and the active cultivation of virtue. The Epicureans favoured *ataraxia*, translated as 'equanimity' or 'imperturbability', which appears a distinctly more timid route of seeking modest pleasures while freeing oneself from fear and bodily pain. Neither school incorporated any idea of sympathy with or consideration for other species – in fact, the opposite. For the Stoics, the concept of *oikeiosis*, that subtle idea of affinity, affiliation or belonging, which formed the basis of their ideas, precluded the inclusion of animals within the sphere of relationships. The absence in animals of *hegemonikon* – a ruling or guiding power which is a precursor to reason as part of the formation of the soul – meant that in their irrationality they could not be regarded as worthy of the extension of fellow-feeling. For both Stoic and Epicurean, the perceived absence of the power of speech in other species further precluded the necessity for giving them any consideration. If an animal was unable to express itself through speech, it was also unable to form any sort of contractual relationship, which placed it outside considerations of justice, or indeed, any consideration at all. Later

Stoic thinkers such as Seneca didn't significantly alter earlier views on relationships with animals.

In the early years of the first millennium, Philo and Alexander, uncle and nephew, both Jewish philosophers living in the Egyptian city of Alexandria, engaged in debate over the treatment of animals. In his treatise *De Animalibus*, Alexander argued that animals have reason and the ability to act morally and to learn and, interestingly for the times, expressed the view that in their comparative weakness animals were like women in that they required protection from men. Philo, more influenced by Stoic thinking, considered it blasphemous even to suggest that animals in their state of unreason might be compared with humans. Origen, a slightly later inhabitant of Alexandria, was a Christian theologian of Stoic persuasion who wrote extensive criticism of the work of the pagan philosopher Celsus, at least in part on the grounds of Celsus's questioning of human superiority over other species. In his beguiling treatise *A True Discourse* – an unsparing critique of both Judaism and Christianity – Celsus writes: 'If men appear to be superior to irrational animals on this account, that they have built cities, and make use of a political constitution . . . this is to say nothing to the purpose, for ants and bees do the same.'

The disagreements, discussions and commentaries which took place between philosophers and theologians such as Philo and Alexander, characterized an intellectually vibrant, disputatious time when new theologies intersected with ancient ideas and influences, many including the fundamentals of human–animal interaction.

The Greek biographer Plutarch, who was born in Greece in 66 CE, and later Porphyry, born in 234 CE in Tyre, were both prolific writers and stern critics of both Epicureans and Stoics. Both wrote extensively about animals, examining the work and thoughts of others and putting forward their own trenchant views on every aspect of human and non-human behaviour. Writing in *De Sollertia Animalium*, 'On the Intelligence of Animals', in 100 CE, Plutarch appeared tender in his observations of other species, of house martins and lions, bees and spiders: 'There is one more reason for admiring spiders ... for there is the fineness of the thread and the evenness of the weaving ... and a tenacity which comes from a viscous substance inconspicuously worked in ... there is the blending of the colours to give it an airy, misty look ...'

Commenting on what he had heard of the behaviour of 'Libyan crows', he described their filling water vessels with pebbles to raise the level of the water until they were able to drink (similar observations in the fables of Aesop may have predated this by several hundred years although there's little evidence that Aesop was a real person or that his fables were the work of a single author. Recent research done at Cambridge University by Nicky Clayton et al. into the cognition levels of corvids provides scientific approbation). In his waspish criticism of Stoicism, Plutarch mocked their parsimony in refusing to have such 'useless and needless' things as 'perfumes and cakes' at their banquets while suggesting that it might be preferable for them to accept that 'untamed and murderous' meat was just as unnecessary.

One of Porphyry's most significant works was *On Abstinence from Animal Food*, a title also translated as 'On Abstinence from Ensouled Food', in which he suggested that being rational, animals are deserving of justice. 'Let us thus exhibit the true view as held by Pythagoras that every soul which participates of sense and memory is rational.' Basing much of his defence of animals on their use of speech, Porphyry suggested that their varied and indicative sounds display their ability to reason and that our own inability to understand them was simply a result of our lack of knowledge. In illustration, he suggested that: 'Arabians understand the language of crows and the Tyrrhenians of eagles,' while he speculated that: 'perhaps all men would understand the language of animals if a dragon were to lick their ears ...' (which may or not be the case). In his long and closely argued theme, he also wrote of animals' ability to remember and to learn: 'They also have memory, which is a most primary thing in the continuation of reasoning and judgment. They likewise have vices, and are envious, though their bad qualities are not so widely extended as in men . . .' Porphyry's most famous work is his treatise *Against the Christians*, and here too he argues his case for consideration for animals. Expressing disapproval of the actions of Christ in causing the demons he exorcized from a man to enter the flock of swine who then tumbled to their deaths from a cliff, Porphyry suggests that Christ might have destroyed the demons without recourse to pig-killing.

While Aristotle assigned us our places on the scale of nature, it was the theologians of Judaism and Christianity who extended the metaphor into the realms of demarcation and hierarchy, encouraging an entirely anthropocentric world view in which humans, made in the image of God, were, and to many still are, both exceptional and superior to every other life form. Conflicting visions of man and his relationship to Earth and to other species appear in Genesis and in later writings of the Pentateuch and Psalms, as well as in a bewildering variety of oral and written law, commentaries and sundry theological writings.

Genesis is contradictory. In the first chapter 'Elohim' – God – creates light and dark, water and earth, plants, stars, sun, moon, fish, birds, sea creatures, land animals, insects and only then man and woman, in his own image, with the purpose of ruling over other creatures. Genesis 1:28 suggests: 'Be fruitful and multiply and fill the earth and subdue it . . .' while in 1:29 and 1:30, God states that both human and animal diet should be: 'every seed-bearing herb, every tree that has seed bearing fruit; it will be yours for food . . . And to all the beasts of the earth and to all the fowl of the heavens and everything that moves upon the earth in which there is a living spirit, every green herb to eat'. In the second chapter, the persona of God is harsher. Yaweh, the Lord God, forms man from dust and woman from bone, although not in his own image. He creates the Garden of Eden but imposes such conditions upon the actions of man that, in the breaching of them, they will bring about the very pain of existence. Later, in Genesis 9:3, the apparent prohibitions of 1:29 and 1:30 are contradicted by the verse that states 'Every creature that lives shall be yours to eat'. (This apparent permission to eat meat has been interpreted by some rabbinic authorities as a way of dealing with mankind's unfortunate propensity for eating the flesh of others, a temporary concession until they could elevate themselves above an inclination towards cannibalism.) In his book *Out of the Earth*, the renowned soil scientist Daniel Hillel points out that the role of humans is different in the two versions of the story of creation. The first version is anthropocentric as man is given permission to dominate the earth while in the second version man is made of soil and by being given the injunction to 'serve and preserve' the Garden of Eden, is assigned the responsibility for the protection of creation. 'Thus, latent in one of the main founts of Western Civilization,' Daniel Hillel writes, 'we have two opposite perceptions of man's destiny.'

In the New Testament, perhaps the best-known references to other species are to birds in the Gospels, in Mark 6:26 and Luke 12:24: 'Consider the ravens: they do not sow or reap and yet God feeds them. How much more valuable are you than the birds?' While at variance on the market value of sparrows – in Matthew 10:29, two sparrows are sold for a penny while in Luke 12:6, the going rate is five for two pennies – the conclusion is the same: if, despite their small worth, God still cares for them, how much more must he care for humans. Elsewhere, in Acts 10:9, Peter describes his experience in the city of Jaffa, where he had a vision of an object resembling a sheet filled with animals and birds being lowered from heaven and the voice of God telling him to kill and eat. Peter told God that he had never eaten anything impure to which God replied that nothing He created was unclean or impure, and after three repetitions of the exchange, the sheet of animals disappeared whence it came.

Both Judaism and Christianity enshrine the belief that the interests of other species are secondary to those of humans, although in Judaism the treatment of animals is codified in a set of rules derived from a variety of sources, principally Leviticus and added to over time, known as *tsa'ar ba'alei chaim*, 'the suffering of animals'. There are both prohibitions and instructions on the way animals should be treated, including detailed ones on the care of domestic and working animals.

One of the great figures of medieval scholarship, Rabbi Moshe ben Maimon, Moses Maimonides was responsible for codifying Jewish law into the Mishneh Torah. Born in 1135 in Spain, Maimonides lived for most of his life in Fustat near Cairo, studying, writing and practising as doctor to the Grand Vizier of Egypt. It is in his book *The Guide for the Perplexed* that he expressed his ideas on the correct treatment of other species, and although he considered humans unique in their ability to comprehend the divine and human duties towards animals to be 'indirect duties' – those which considered the effects of the treatment of animals as they might affect humans – he believed animals to be no less capable of sentience and imagination than humans: 'Why should God select mankind as the object of His special Providence, and not other living beings?' He wrote: 'There is a rule laid down by our sages that is directly prohibited in Law to cause pain to an animal' and 'We should not kill animals for the purpose of practising cruelty, or for the purpose of play.' As a doctor, he recommended the eating of meat for the benefit of health, but reiterated the laws of *tsa'ar ba'alei chaim* prohibiting cruelty in slaughter, the cutting off flesh from live animals, or the causing of grief to birds or animals, strictures that had little influence on the thinking of many later philosophers. Islamic law is similar to Jewish law in many respects, with rules to protect and safeguard the lives and well-being of other creatures. The Qur'an states: 'There is not an animal that lives on the earth, or a being that flies on its wings, but they form communities like you' and 'And the earth, He has assigned it to all living beings.'

Why, of all possible modes of thought, did the ones that would encourage the greatest licence and cruelty to other species prevail? Why were the texts that engender a view of the desirability of a hierarchical, exploitative relationship between humans and animals the ones to gain ascendancy? Why was there so little attention paid in Western thinking to the more questioning and humane voices that sound throughout the worlds of Greek thought, of the Pentateuch and other books of the Bible, in Ecclesiastes: 'For that which befalleth the sons of men befalleth beasts, even one thing befalleth them, As one doeth, so doeth the other; yea they all have one breath; so that a man hath no pre-eminence over a beast, for all is vanity' or the Psalmists, in particular in that greatest of poetic praise-songs to the life of Earth, Psalm 104 '... where the birds make their nests, as for the stork, the high junipers are her house. The lofty mountains are a refuge for the ibex and the rocks a shelter for the hyraxes ...' with its overlaps and whisperings from earlier stories and the creation-myth accounts of the beginnings of life?

One of those whose thinking would most influence later ideas on how humans might treat animals, was Aurelius Augustinus, Bishop of Hippo, later St Augustine, whose view of other species was entirely Stoic: 'If when we say, Thou shalt not kill, we do not understand this of the plants since they have no sensation, nor of the irrational animals that fly, swim, walk or creep since they are dissociated from us by their want of reason and are therefore by just appointment of the Creator, subjected to us to kill or keep alive for our own uses.' In chapter 17 of *The Catholic and Manichaean Ways of Life*, Augustine wrote: 'We can perceive by their cries that animals die in pain, although we make little of this since the beast, lacking a rational soul, is not related to us by common nature.' Of other species he said: '*Non agunt sed magis aguntur*' ('they do not act, they are acted upon').

Possibly the most important figure in reinforcing these ideas, one whose influence appears relatively undiluted by time, is the Italian

Dominican St Thomas Aquinas. Born in central Italy in 1225, Thomas, who was known as 'the angelic doctor', was author of one of the greatest works of Catholic scholarship, the *Summa Theologica*, written during the final nine years of his life and still one of the most important texts in Catholic teaching. Despite being influenced by Maimonides, particularly in the formation of the third of his five proofs of God's existence and by *The Guide for the Perplexed*, Thomas's beliefs on the nature of other species differed greatly from those of Maimonides. Based on Aristotle's hierarchy of natural life and his own formulation of the components of relevant souls and the first chapter of Genesis, their conviction was that being made in the image of God conferred superiority upon humans. (Given that he believed women to be of lesser intellect and merely 'defective' men, a mysterious situation possibly brought about by the blowing of moist south winds during their conception, one might be less than optimistic about his views on animals.) In lacking rationality, he thought, animals 'are ordered to man's use ... without any injustice, either by killing them or employing them in any other way' and that 'He who kills another's ox sins, not through killing the ox, but through injuring another man in his property.'

In March 1967, a paper written by the American historian Lynn T. White was published in the journal *Science*. Based on a lecture he had given the previous December to the American Association for the Advancement of Science, it was entitled 'The Historical Roots of Our Ecological Crisis'. In it, he traced the history and effects of Western – primarily Christian – thinking on the development of the very technologies which he asserted were leading to ecological disaster through the use of fossil fuels, population explosion, 'planless urbanism' and excessive waste production. In its belief that God's will is for man to exploit nature, Christianity is, according to White, 'the most anthropocentric religion the world has seen'.

Far from being a dogged and contentious critic of the doctrines of

Christianity, White, a professor of history and scholar of medieval technology, was deeply concerned about the fate of other species, wishing to see them incorporated into a 'spiritual democracy of all God's creatures'. Describing himself as a churchman, he nonetheless placed the responsibility for what he saw as an impending ecological crisis firmly at the hands of Christian thinkers and theologians over the centuries whose influence he asserted is as profound as it has been since its inception. 'The victory of Christianity over paganism', White wrote, 'was the greatest psychic revolution in the history of our culture . . . We continue today to live, as we have lived for about 1700 years, very largely in the context of Christian axioms.' Only when we reject the idea that nature exists only to serve man, White wrote fifty years ago, will we avert the worst of future planetary harm. White's ideas have been the subject of criticism, discussion and re-evaluation in the years since their airing, much by Christian thinkers and commentators. In his superb book *Man and the Natural World*, the historian Keith Thomas points out that White has been criticized on the grounds that many religions and societies have found it expedient to exploit the world's resources to the extreme detriment of other species, including those whose social and religious foundations might have suggested they do otherwise by scripture and outlook. Many in the field of ecology regard his work as still being both relevant and important. Writing in the journal *Nature Ecology and Evolution* in December 2016, Michael Paul Nelson and Thomas Sauer write of White's assertion that 'the man-nature dualism is deep-rooted in us' and repeat White's warning that the ecological crises will only worsen until we reject the idea that nature has no reason to exist except for the use of man.

The axioms to which White refers form a long and continuing line, one broken only occasionally by those whose views about how we should live in the world were both openly expressed and at variance with the orthodox, Stoic-derived teachings of the church. St Francis of Assisi included all species in his vision of a single indivisible life on earth. He is said to have preached to birds and gained the trust of a marauding wolf. In writing of him as 'the greatest spiritual revolutionary in Western History', Lynn T. White admired Francis's attempts to alter the prevailing view of superior man and subservient creature but acknowledged that it was an attempt that failed, crushed by the very orthodoxy that White believed has brought about the crises of ecology.

There are plenty of accounts of saints having shown compassion towards other creatures but, as with the influence of St Francis himself, they seem little more than distractions in the continuing process of the distancing of humans from other species. St Francis of Paola may well have resurrected his pet trout, St Isidore may have shared his food with hungry birds and St Martin de Porres exercised kindness towards mice and rats, but for all the many saintly instances of kindnesses, mutual understanding, sympathy and aid, they seem to have had little effect in altering the prevailing certainties of the inferiority of others which encouraged the belief that the human – in particular the human male – was beyond the need for restraint in their behaviour towards any other creature. The expression of sympathy towards animals didn't – as it still doesn't – necessarily suggest an all-encompassing love for every living being. Martin Luther's expression of fondness for his dog: 'Be comforted little dog; thou too in Resurrection shall have a little golden tail' may be a charming sentiment but hardly takes the mind from encouragement to murder and the other manifestations of the virulence of his anti-Semitism.

In 1571, now retired from the onerous duties of public service and settled into the quiet of one of the tower rooms of his family chateau in the Dordogne, Michel de Montaigne, a French nobleman of unusual education and original mind, began to record his thoughts in a long series of remarkable and enduringly popular essays in which he displayed rare empathy for the lives and deaths of other creatures, and indeed other humans. In his essay 'On Cruelty' he wrote:

... I have not even been able to witness without displeasure an innocent defenceless beast which has done us no harm being hunted to the kill ... throwing itself on our mercy which it implores with its tears ...

I hardly ever catch a beast alive without restoring it to its fields. Pythagoras used to do much the same, buying their catches from anglers and fowlers ...

Watching animals playing together and cuddling each other is nobody's sport: everyone's sport is to watch them tearing each other apart and wrenching off their limbs.

These were all uncommon sentiments for his times. (Despite them, he did still go hunting.) In his longest essay 'An Apology for Raymond Sebond', described by the French philosopher Jacques Derrida as 'one of the greatest pre- or anti-Cartesian texts on the animal that exists', Montaigne wrote extensively of other species, of elephants, eels, bees, goats, oxen, magpies and others, praising their qualities and abilities. After a beautiful passage describing the return of swallows in spring and the construction of a spider's web, he wrote:

We are perfectly able to realise how superior they are to us in most of their works and how weak our artistic skills are when it comes to imitating them. Our works are coarser, and yet we are aware of the faculties we use to construct them: our souls use all their powers when doing so. Why do we not consider that the same applies to animals? Why do we attribute to some sort of slavish natural inclination works that surpass all that we can do by nature or by art?

His musings on his cat: 'When I play with my cat, how do I know that she is not passing time with me rather than I with her?'

are reminiscent of Zhuangzi, the Taoist philosopher of the fourth century BCE, wondering if, when he dreams of being a butterfly, he is Zhuangzi dreaming of being a butterfly or a butterfly dreaming of being Zhuangzi.

It might have been anticipated that these particular thoughts of Montaigne's would not be embraced by later philosophers, René Descartes and his follower Nicolas Malebranche among them. In a letter to the Marquis of Newcastle on 23 November 1646, Descartes wrote:

> As regards the understanding or thought ascribed by Montaigne and some others to animals, I cannot share their opinion . . . if they thought in the same way that we do, they would have an immortal soul as we do; which is not plausible, because there is no reason why we should believe it of some animals, without believing it of all, and that there are many too imperfect for it to be possible to believe in their case, such as oysters, sponges and so forth.

Descartes's beliefs on the nature of other species are derived from his speculations on the relationship between mind and matter in his division of all substance into *res extensa*, bodily or corporeal substance and *res cogitans*, mind or soul. In *Discourse on Method*, published in 1637, Descartes wrote: 'And here I specially stayed to show that, were there such machines exactly resembling in organs and outward form an ape or any other irrational animal, we could have no means of knowing that they were in any respect of a different nature from these animals,' the basis of the term commonly attributed to him, *bête machine*, 'animal machine'. Later he wrote:

> I here entered, in conclusion, upon the subject of the soul at con-siderable length, because it is of the greatest moment: for after the error of those who deny the existence of God . . . there is none that

is more powerful in leading feeble minds astray from the straight path of virtue than the supposition that the soul of the brutes is of the same nature with our own; and that consequently that after this life we have nothing to hope for or fear, more than flies and ants . . .

There are those who think it inaccurate to interpret Descartes's beliefs as meaning that he denied all thought and feeling to animals, or the possibility of their experiencing suffering. In a paper 'A Brute to the Brutes? Descartes' Treatment of Animals', published in *Philosophy* in 1978, the Cartesian philosopher John Cottingham suggested that, rather than Descartes being quite as convinced of the idea that animals are no more than machines, the appearance of his being so is caused by contradictions in his metaphysical theories as they deal with the strict boundaries between *res extensa* and *res cogitans*. Ascribing this to a 'certain fuzziness' in Descartes's thinking, Professor Cottingham suggested that his intentions were less 'beastly' to other species than previously supposed. This conclusion might seem at variance with Descartes's own apparently enthusiastic ventures into vivisection. Although the frequently quoted story that Descartes nailed his wife's dog to a board for the purpose is untrue, this is not because he wasn't given to episodes of vivisection but because he never married. (He is reported to have had a pet dog of whom he was fond, one Monsieur Grat, who, fortunately enough for him, was sent for stud purposes to a friend's dog in Paris.)

Whatever Descartes believed, his follower, the French theologian and priest Nicolas Malebranche, appears to have adopted enthusiastically the wholly negative view of animal sentience: 'Thus in animals there is neither intelligence nor soul in the sense as is ordinarily meant. They eat without pleasure, cry without pain, grow without knowing this. They have no desires and no knowledge,' and as if to display the rigour of his Cartesian orthodoxy, he's reputed to have kicked a pregnant bitch deliberately, which if true demonstrates only that you can be both Oratorian priest and lout at the same time. In a critique of Montaigne's life and work, he wrote archly, 'he worked hard to give himself the air of a gentleman but he did not work to give himself a precise mind', deriding him for referring to other species as, 'our fellow brethren and our companions'.

The essence of unchangeable certainty lies along the same fixed historic line: 'I am very much obliged to you,' Descartes wrote in a letter to Father Denis Mesland, a Jesuit living in Leiden in May 1644, 'for pointing out the passages in St Augustine that might serve as authority for my views ... I am highly satisfied that my thoughts should agree with those of so holy and distinguished a person ...' In *The Passions of the Soul*, a summation of the ideas Descartes had been exchanging in a fascinating correspondence with the erudite and incisive Princess Elisabeth of Bohemia, he expounded the nature of the relationship between the 'passions' and the soul. Identifying six 'primitive' passions: wonder, love, hatred, desire, joy and sadness, he sought both to define and expand on them and to describe the body's interactions with the soul: '... all animals without reason conduct their lives purely by corporeal movements similar to those which, in us, typically result from the passions ...' he wrote.

The line tightens, connects us back, ties us to the past. It's straight, direct, unyielding, one which continues to demarcate us, this from that. In his book *Lines*, the anthropologist Tim Ingold explores the lines which lead, connect and direct our lives in art, movement, music, writing and every human endeavour. He suggests that in Western societies, straight lines have come to denote the discontinuities between ways of thinking and perceiving, representing 'the triumph of the rational' which divide culture and intellect from intuition and nature.

Not long ago, half-listening to the radio, I heard two academic philosophers discussing what differentiates us from other species. I was busy cleaning the birds' houses, and allowed their voices to fade in and out like signals from an approximate planet. I don't know what took my attention back but I was in time to hear one of the participants say that the difference between humans and other species is that despite our all being 'biological creatures', other species need nothing beyond the fulfilment of physical needs. He used horses as his example – if they have food and 'a nice field', he said, they're happy and have no further requirements, unlike humans who have a desire for perfection, 'transcendent urges', souls, and an understanding of the finitude of time.

Cleaning a bird's house is more than it seems. It involves an acknowledgement of ownership and territory, and a careful weighing of the balance between respecting the innate corvid practice of hiding things purposefully for later with the finer aspects of hygiene. I always know that when the work is completed, both rook and crow will check it with the air of a stern employer who, by character or experience, distrusts the staff. After I'd heard the philosophers' discussion, their words wouldn't go away. I listened to it again and for the rest of that day and every day since, I've wondered how anyone can still hold such reductive views of other species.

The subject of their discussion – how humans differ from other species, perfectly characterizes the idea known as 'human exceptionalism', the enduring belief that humans are not only different from other species but superior, God-chosen, apart. Based on the fundamentalist interpretation of Genesis and closely allied to ideas about the possession of a soul, this particular certainty has been challenged by evolution, rejected by many and now is widely regarded among those concerned with the fate of earth as one of the means by which the continuing ruination of the natural world has been facilitated but it remains, stubbornly, irrationally, immovably here. The longstanding effects of our dualist thinking, 'them and us', souls or no souls, complex brains or not, 'reason' or not, are embedded in our language and ideas, enduring and pervasive. We still act towards other species as we do because, even if we don't we believe they lack a soul, we really do believe ourselves to be different and exceptional. We still deny to others appreciation of the elements of life which animate us all to sing, love, fear, swim, fly or consider, to be part of our family or to extend ourselves towards others of our own species or not. Beyond the confines of the theological or theurgical, the soul, whatever it is or can be called, still seems there, invisibly intrusive, defining the boundaries of everything that separates species from one another.

The philosopher Mary Midgley writes, 'The first thing we have to do is to get ourselves in proportion – to see through our current absurd over-estimate of human separateness and superiority.' With the diminution of religion as a force, Midgley writes, 'No higher power can now interfere with us. This attitude ... has become, not just anthropocentric but effectively anthropolatrous, self-worshipping. It sees us as omnipotent.' Her view is reinforced by the constant and recurring claims made by those who believe humans to be exceptional, among them that the finest exponents of one or other human art form, in some way exemplify the attributes of humanity as a whole. Mozart, Bach and Shakespeare all seem to suffer particularly from being gathered into the malign fold of 'exceptionalism', referred to often and quoted as proof of human superiority. Because they, as humans, were exceptional, the reasoning proceeds, we, as humans must all be too, gilded by the mere fact of our species, with the radiance of their achievement. We may be, but in comparing ourselves to rare geniuses most of us fail to measure up. If we do share traits with geniuses, they may not be ones that suggest any universal ability to compose like Bach or write like Shakespeare. The implication is too, that every other species aspires to write or compose or to achieve the wonders of humanity but, in not being fortunate enough to be human, they are doomed never to be able to reach the glory of our ever-ascending heights.

The primatologist Frans de Waal writes critically and engagingly about 'human exceptionalism' in his book *Are We Smart Enough to Know How Smart Animals Are?*, drawing on his vast experience of working with and observing the behaviour of primates to evaluate their capabilities, and ours. In order to allow for work on how cognition operates across nature, he suggests that space might be found to pursue it if there were, 'a moratorium on human uniqueness claims'.

The belief in human exceptionalism might be challenged if equal regard and time were paid to avian exceptionalism, piscine exceptionalism, the unique qualities and exceptionalism of every species. As humans, our tragedy may be that we can judge only by our own methods, our own ideas, preconceptions and prejudices, caged within the bounds of our all too limiting human view, our cosmic, wanton solipsism.

It is not very long ago that the sphere of human exceptionalism would have excluded women and people of differing ethnic or religious groups from the divine forum of the blessed, the self-selected elite of Western societies, white men. The idea has increasingly entrenched itself as a defence of political and economic interests, espoused by those who are afraid that if other species are given any consideration, however small, their lucrative commercial activities might be curtailed and their freedom to mine, chop down or burn rainforests, build roads and pipelines, might be threatened. A leading American defender of the idea of human exceptionalism believes that the 'demotion' of humans to 'just another forest animal' will lead us to behave like forest animals which, given the nature of well-functioning forest environments and their many mutually beneficial and cooperative species, can surely only be a good thing. Claiming that humans are good at what we're good at may be uncontentious, but it fails to acknowledge the very many things at which we're lousy, such as recognizing the speed at which we're destroying the planet on which we currently live.

In *The Soul of an Octopus*, Sy Montgomery's intriguing, respectful examination of the life of these strangest, most wonderful of underappreciated creatures, she muses on the nature of the soul and what it is. She speculates on definitions and parameters, asking if it is emotion or consciousness, 'the fingerprint of God' or 'the indwelling consciousness that watches the mind come and go'. Uncertain if it's all of these or none, she concludes: 'I am certain of one thing ... If I have a soul – and I think I do – an octopus has a soul too.' I've reached that conclusion – if I have a soul, so do my birds.

On my phone, a photo appears of an abandoned cat, recently rescued. The creature, a handsome Orlando-style marmalade cat, has wet and reddened eyes. 'Can anyone doubt that animals feel things and have souls?' the person who posts the photo asks. It's difficult to interpret the feelings of another, but when I look at this cat I can't fail to know that he suffers but whether it's the result of a malady of body or soul, I can't begin to know. In my uncertainty, all I can do is trace the lines, run after the scattered beads, ideas and questions of the hierarchy, gender, power and cruelty enduring in the things we're so reluctant to relinquish – our food, clothes, the pleasure some derive from the killing of animals for sport and wonder if, for a lack of appreciation of what might be other's souls, we seem to have lost a part of what might be our own.

4

Blood

A flat, grey day in November and Camden unlovely on a damp and colourless Sunday morning. As I'm walking from the depth of the tunnels of the Tube, people seem to flow upwards past me like geysers of movement bubbling up from caverns of darkness to flood out into the light of the street, turning as a tide towards the crowds in Camden Market. I walk up Parkway towards the Jewish Museum. Years ago, I lived nearby but nothing makes me want to remember, nothing makes me want to re-inhabit all the layers of image and memory from another, distant life. I walk past the premises which once was the pet shop where our first rat Rupert was bought and later Marley, the sun conure. It's a cafe now although the famous sign's been preserved: MONKEYS, PALMERS, TALKING PARROTS, REGENT PET STORES, PALMERS, NATURALISTS.

It's early when I get to the Jewish Museum and the place is almost empty. I have visited the museum many times before so the permanent displays are familiar; the artefacts from Jewish life of the past, the objects my great-grandparents' generation brought with them from Eastern Europe or lost there, things made or bought, the candlesticks

and textiles and Kiddush cups, all versions of ones I have and sometimes use, the inherited silver *havdalah* for use in the ceremony to
differentiate the Sabbath from its ending with the sighting of three
stars in the sky, the hand-embroidered bread covers now worn and
frayed, the one with the words spelled out in threads in Hebrew
'Remember the Sabbath day, to keep it Holy' and the unravelling
clover leaves round its edge. Wandering through the museum, I reflect
that I could put on a small but comprehensive exhibition of my own,
an exclusive – Scottish Judaica – should the mood ever seize me, which
I can say confidently, it won't.

The exhibition I'm here to see is about blood. A substance of
particular religious and historic significance in Judaism, it is to be
contemplated on the top floor: our relationship to blood, our own and
others', the determinant of us all in the actuality and in the way we use
the word, of act, thought or lineage, in blood lines, royal blood, blue
blood, in blood feuds, blood sacrifice, blood libels – expressions we
use to try to differentiate and elevate and separate – this fluid that is,
if not in its cells or colour or type then in its almost universal import,
the shared components of our physical being, the blood we may share
or shed or consume, and the ways in which we do it. *Ha dam, who ha
nefesh* it says definitively in Deuteronomy 12.23, 'blood is the soul . . .'

As I examine the cases and displays of the exhibition – its beautiful,
thoughtful evocations of our lives seen through this single notion of
blood – one artefact keeps drawing me back: a set of killing knives
lying in neat sheaths of wood engraved with the names and images
of the creatures for whom they were intended: cow, chicken, sheep.
Utilitarian and worn, they seem to hint at a life of simplicity. They,
like my family, came from Lithuania over a hundred years ago. These
knives used in the slaughter of animals for meat take me to everything
connected with what we eat and why and the ways we do it. They
linger in my mind after the visit, as I walk back out into the November
air, connecting me to the religious and philosophical thinking of

millennia – where do the boundaries lie for what I might or might not do to other creatures? What allows me to do it? Where does the killing of another creature and the consumption of its flesh place a human in relation to the world of other living beings? How did we begin to reflect on the idea of the taking of others' lives for food? The leaflet I brought back from the exhibition has a photo of the knives and bears the title: 'Blood, Uniting and Dividing'.

After I get back from London, I take my prayer book out, the one I had at school, my *Siddur*, worn more with time than overuse. I look at it before the festivals which I celebrate, sometimes, with the enthusiastic but ameliorated dedication of the cultural Jew, when I need to remind myself of the appropriate blessing or passage to be read, to check a quotation from a psalm or prayer or look up a reference. On the flyleaf, at what in an English-language book would be the back, is my name in my father's hand. This one's configured as Hebrew books are although it's a parallel text. I open it at 'Additional Service for the New Moon' – the moon being significant in a lunar calendar, and on page 225b, read:

> And in the beginnings of your months ye shall offer a burnt offering unto the Lord; two young bullocks and one ram, seven he-rams of the first year without blemish. And their meal offering and their drink offerings as hath been ordained, three parts of an ephah of fine flour for each bullock, two tenths parts for the ram, one tenth for each lamb, and wine according to the drink offering of each, a goat for atonement, and two daily offerings according to their institution.

(An ephah, it seems, is equal to half a gallon dry measure.) I have lived with this book and the influence of its contents all my life without really thinking about what this opening line means and that 'ye shall offer a burnt offering' is instruction to sacrifice an animal, or would be if sacrifice in Judaism hadn't ended with the destruction of the Second Temple in Jerusalem in 70 CE. It would all seem nothing more than a weirdly horrible ancient practice if it didn't tell far more about our relationship to other species than of the simple shedding of their blood.

Sacrifice has been carried out in almost every time, place and society – killing as act of propitiation, divination, expiation or supplication, as a display of religiosity, the exercise of power or in the worship or placation of many and various spirits, deities, overlords or unseen forces. It has served as a means of legitimizing the forbidden, of reinforcing kinship, particularly male kinship, and of encouraging reciprocity and exchange. Sacrificial remains of both humans and animals scatter the world's archaeological sites from South America to China, from Africa to Central Asia. Animal sacrifice took place in all the early civilizations of the Middle East, in Egypt and Greece, Rome and in the civilizations of the Canaanites, Israelites and later, Jews. Archaeological evidence from a site in central Palestine dating from the 5,000-year-old Canaanite civilization shows that animals were being imported from the Nile Valley for sacrificial purposes while further north at Tel Megiddo, now in northern Israel, at much the same time, sacrifices of sheep, goats and cattle were being made to still unknown gods, their bones found deliberately laid out in long corridors in the Great Temple, apparently as part of the ceremonials of sacrifice. In both Greece and Rome, sacrifice was an established practice, in some ways similar to Temple sacrifice except in the vital respect that in the course of time Temple sacrifices were made only to a single deity.

It is disquieting, dealing with elements of our earlier selves, examining deeds remote in time and imagination, abhorrent in execution, obscure in intent. Reading the best-known text on animal sacrifice – Leviticus, the third book of the Pentateuch – does not help explain any of it, perhaps because the earliest of its intended readers already had some acquaintance with its arcane rituals and meanings. Described as 'instructions for priests', Leviticus offers a grim, often tedious list of instructions on the correct way to carry out sacrifice and on all aspects of the lives of Jews, from ones related to health, sex, human interactions and agriculture, to which creatures might be eaten and which not, the good sense of which seems borne out by the fact that until today, not many make a habit of eating hoopoes or bats.. From its dramatic first word: *Vayikra*, 'He called . . .', descriptions are livid on the page: 'And he shall slaughter the young bull before the Lord, and Aaron's descendants . . . shall bring the blood and dash the blood upon the altar . . .' but shocking though they appear now, Leviticus's injunctions as to which creatures should or shouldn't be killed for food and how they should be eaten, may have been among the first codified rules on our using other beings for food, however complex and layered, however alien it all seems now.

Once an act that might be undertaken by individuals anywhere, random sacrifice was stopped following the stern prohibition on pagan practices carried out by King Josiah during his reign from 640–609 BCE. After that, animal sacrifice was to be carried out only by priests in Solomon's Temple which, from the time of its construction in Jerusalem around 950 BCE, until its eventual destruction by the Romans in 70 CE, was, officially at least, the site of all Jewish sacrifice. The laws of sacrifice, though crazily intricate and certainly not always strictly observed, were clearly defined. Human sacrifice was prohibited and animal sacrifice was limited to those animals permitted under the orally transmitted Mosaic law for the purpose of atonements of various kinds, such as the *oleh*, the burnt offering, given in its entirety to God, or those of thanksgiving. Grain and the first fruit of the season could be offered. The meat of sacrifice was shared between the priests and the bringers of the animals. No fat could be consumed and no blood. ('Be resolute in not eating blood,' says Deuteronomy 12: 23–24, 'for blood is the soul and you will not eat the soul with the meat. Do not eat it – spill it on the earth like water.') The only animals who could be sacrificed were domestic. Doves could be sacrificed too so that the poor could contribute and eat – an important provision in a society in which meat-eating was rare and sacrificial animals precious to their owners. The very poor weren't required to provide a creature – they could bring one tenth of an ephah of flour as an offering.

Sacrifice ended with the destruction of the Second Temple in Jerusalem in 70 CE. However terrible these acts seem now, they weren't (or shouldn't have been) carried out without thought. Taking an animal's life was regarded seriously. The placing of the hands on the creature before sacrifice was deemed the symbolic identification of human with animal, a moment of contemplation, transformation and repentance, what the French anthropologist Marcel Mauss considered to be both a religious and magical merging of sacrificer and sacrificed. The laws on the correct treatment of animals prohibited the slaughter of a mother and offspring on the same day and should have ensured that an animal's pedigree had to be known. This may have constituted a stage in progress from the previous episodes of human sacrifice, wanton killing, animal torture, sadistic display and excess that occurred throughout the ancient world of the Mediterranean.

Trying to understand earlier versions of ourselves is a long trail through social, religious and political development and change, only to arrive back at the questions still to be asked, the one I asked the spider on an October afternoon: *How, in the light of our shared past should I behave towards you and others?* We still ask questions daily about the spilling of blood and though the spider and I might not be precisely the same in blood – hers is haemolymph, a clear, pale bluish liquid composed of haemocyanin, nitrogen, carbon, hydrogen and oxidized copper while mine's just the ordinary human stuff – our mutual reliance upon it is indivisible.

The ways we discuss how and why we kill other creatures for food often seem like recent concerns but they may always have seemed that way. In Book 15 of *Metamorphoses*, Ovid wrote of Pythagoras: '... He was the first to say that animal food should not be eaten ...

there is corn for you, apples as whose weight bears down the bending branches . . . she offers for your tables, Food that requires no bloodshed and no slaughter,' making it clear that, like the matter of the possession of a soul, it was a topic central to contemporary thought.

In *On Abstinence from Animal Food*, Porphyry began by addressing himself to a friend who he'd been told – clearly through some gossipy vegetarian grapevine – had abandoned the practice of vegetarianism. It's the kind of letter nobody enjoys receiving, a 'more-in-sorrow-than-in-anger', 'I always thought you were better than that' kind of letter. The unfortunate recipient, Firmus, perpetrator of this shameful dietary backsliding can't have been in any doubt of the extent of his folly: 'when I was informed by certain persons that you have even employed arguments against those who abstained from animal food, I not only pitied, but was indignant with you . . . you have deceived yourself and have endeavoured to subvert a dogma which is both ancient and dear to the gods . . .'

In her excellent biography *Seneca: A Life*, Emily Wilson describes Seneca's short experiment in vegetarianism. He embarked on it under the influence of his teacher Sotion but abandoned it after a year on the insistence of his father who was concerned by society's increasing hostility towards 'foreign rites', which may or may not have included vegetarianism. The 'foreign rites' in question seem to have included Jewish and Egyptian dietary habits, which aroused such ill-feeling that they were punishable by expulsion and transportation. Under his father's exhortations, Seneca returned apparently without qualm to 'dining more comfortably'. He did however, dislike the fashionable sight of fellow-diners engaging in a peculiarly revolting practice whereby a red mullet, a 'surmullet', was brought to the table alive in a receptacle of glass where they would watch it die before eating it, although his objection appears to have stemmed more from a proper Stoic disgust for excess than any fastidiousness about watching creatures suffer in death.

Plutarch too, a particularly passionate advocate of vegetarianism, expressed himself with often touching percipience in his essays 'De Sollertia Animalium' and 'De Esu Carnium'. In the latter, he wrote of his wonder at how it might be possible for man: 'to reach to his lips the flesh of a dead animal, and having set before people courses of ghastly corpses and ghosts, could give those parts the names of meat and victuals, that but a little before lowed, cried, moved and saw', wondering too how it's possible to 'pollute' oneself with meat when an abundance of other foodstuffs exists. His criticism of the way meat's prepared to make it edible has an oddly contemporary ring: 'to this same flesh, mixing oil, wine, honey, pickle and vinegar, with Syrian and Arabian spices, as though we really meant to embalm it.' Convinced that meat-eating was injurious not only to body and mind, he asserted that as licence to further excess it also was injurious to the moral well-being of man.

Even beyond considerations such as suffering or souls, the implications of eating the flesh of other creatures is profound. Is it 'natural' for humans? Is it necessary? Are the processes of its production cruel? Are we part of the natural world of predation and consumption, part of a cycle of being from which we can extricate ourselves only with difficulty and the danger of distancing ourselves still further from the processes of nature?

These questions have persisted through the centuries. Keith Thomas wrote in *Man and the Natural World* of the disquiet and distaste felt by many in the seventeenth century about the wholesale slaughter of animals for food, among them Dryden, Aphra Behn and John Evelyn. The very arguments used today in support of vegetarianism, he suggested, were discussed then – the deleterious effects of meat-eating on health and character, the cruelty involved and unexpectedly – for the idea seems a purely contemporary concern – the inefficient and uneconomic use of land for stock breeding rather than arable farming. In addition to the vegetarians of the classical world, there have been those throughout time who we believe to have been vegetarian (although being absolutely certain of the secrets of another's diet is always going to be difficult). Many early Christians such as Origen and St John Chrysostom were said to be vegetarian, as were Leonardo da Vinci, Percy Bysshe Shelley, Mary Shelley, Kafka and Tolstoy and many more. In her essay 'Sister Turtle' the American poet Mary Oliver considers the question. Having eaten very little meat for years, she talks of an occasional craving, describing the human desire to eat meat as 'this other-creature-consuming appetite', what she describes as a 'miraculous interchange' in the way in which one creature's flesh nurtures that of another as an act of continuity, invoking the religious dimensions of body and blood. The ambiguity she talks of in her anxiety for the about-to-be slaughtered lamb and for her own body and soul ('that worrier' as she calls the soul) characterizes elements of the uneasy way we – in the West at least – live with ourselves and others, the irresolution with which, increasingly, we consider the sources of our food and the means of its production.

Whether by eating them or not, the human diet has always involved animals. The evidence is there, lying on the fringes of many coastal cities, including New York, built on the remains of shell middens – the scattered remains of oysters, mussels and other shelled sea-creatures or in the signs of early meat-eating, widely detected from tooth morphology and butchery marks on bones. Through their examination of the detritus of the early world, archaeologists, palaeoethnobiologists, osteologists and others have revealed at least some of the secrets of the human table (or whatever piece of similar furniture might have been available).

Obtaining meat has always involved heavy energy expenditure although it is still uncertain whether, for the early diet, it was obtained through hunting or scavenging. According to many studies, the belief that the diet of our Palaeolithic forebears was a single, homogenous, heavily meat-centred one is wrong if only because, given the vast territorial reaches and hugely variable climatic conditions over a very extended period, no single diet can be put forward as the one to which humans 'were evolved' to consume.

The diet of early humans, according to most sources, was composed of vegetable matter, fruits, nuts, insects, birds and small creatures and only by the domestication of animals was the availability of meat for human consumption no longer reliant on the heavy expenditure of energy involved in hunting. Even after domestication, meat-eating appears to have been limited for practical and economic reasons in the ancient world of the Near and Middle East, as it has generally been for most people everywhere until recently. Those creatures of poorer land, goats and sheep, were the most common animals in the ancient world but their meat wasn't eaten daily. Traceable diets seem remarkably similar to those eaten in the region now, most calories being obtained from lentils, cereals such as barley, kamut and emmer, vegetables, grapes, olive oil, and from the dates, honey and figs which appear so frequently in biblical texts.

How we obtain the flesh of the animals we eat, how we raise and breed and treat and kill them are all closely related to the way we look at our own species and the extent to which we give ourselves the privileges apparently bestowed by history and by God. For defenders of the idea of 'human exceptionalism' and those who believe that animals are there for the benefit of man, the question may be simple. For those less certain about the divinely inspired supremacy of *Homo sapiens*, it's not.

In 1980, John Berger wrote and narrated the script for *Parting Shots from Animals*, a film made for the BBC television Omnibus arts series. Written from the viewpoint of the animals portrayed, it provided insightful commentary on human–animal interaction:

A very long time ago you worshipped us ... Then 300 years ago, Descartes divided every creature in the world between those who had souls and those who didn't. Men had souls, animals didn't. We were simply machines. We haunted him a bit. But then later you invented real machines to replace us. What the silicone chip is for you, the petrol machine was for us.

This metaphoric machine with its central image of usage and commodification, extended in reach over time to become the means by which animals were not only replaced by engines but were themselves transformed into products processed by machines as the vast, unfeeling apparatus of technology overwhelmed the species we have domesticated for our own purposes. Over the past century or so, the industrialized production of animals for food has subsumed pastoral agriculture into enormous international enterprises on a scale beyond imagining. Questions such as the possession of feelings and minds, never mind souls, by other species seem to have been sucked into a whirlwind of swift and powerful rapacity.

It is decades now since someone in America's Environmental Protection Agency thought up the terms 'Animal Feeding Operation' and 'Concentrated Animal Feeding Operations', more commonly known by their detached, invidious acronyms, AFOs and CAFOs. The Natural Resources Conservation Centre, a United States Department of Agriculture agency provides a definition of AFOs: 'agricultural enterprises where animals are kept and raised in confined situations. AFOs congregate animals, feed, manure and urine, dead animals and production on a small land area. Feed is brought to the animals rather than the animals grazing or otherwise seeking feed in pasture, fields or on rangeland. A CAFO is . . . a large concentrated AFO.' CAFOs are often referred to as 'factory farms' or 'mega farms' although the bleak word 'feedlots' is more apt. The letters and definitions convey none of the actuality of what a CAFO is, of the numbers of creatures involved, the nature of their lives, the social and environmental consequences of their presence, the frank and inescapable realities of both idea and practice. To be designated a CAFO in the United States the numbers of animals raised in one facility must be 125,000 broiler chickens,

82,000 laying hens, 2,500 pigs, 700 dairy cattle or 1,000 beef cattle. Once a phenomenon solely of American agriculture, the CAFO has spread worldwide. In Britain where there are more than 800 CAFOs, there are as yet no legal definitions of the term. Installations designated 'intensive' farms raise at least 40,000 chickens, 2,000 pigs and 750 breeding sows and often, many more. In all these places, birds and animals are kept indoors, in factory buildings, sheds, confined in small spaces for the duration of their invariably short lives, denied every expression of the social, maternal, sexual or environmental behaviour natural to their species, prone to the many diseases of creatures kept in such circumstances, genetically and behaviourally manipulated to maximize their usefulness as food at the cheapest cost to the producer.

The natural lifespan of a cow is around twenty-five years but most will be killed before they're five, beef cattle usually at around eighteen months. The natural lifespan of a chicken is seven to eight years although some have been known to live for twenty-five years. They are killed at six months. Sheep, able to live for fifteen years, are killed at six years. Pigs, with a lifespan of twenty years or so will be killed at no more than three years old. Geese, turkeys and ducks, all capable of living for ten to fifteen years, will be killed within weeks, a maximum of twenty-three or twenty-four. Those who raise animals this way defend the practices on grounds of what often seem like the contradictory bases of better 'animal welfare' and the necessity to produce more and cheaper food.

It may be easier to believe that most birds and animals eaten come from 'family farms', from open country and pastures and wind-blown grazing, which of course some do. But in Britain, of the billion chickens raised in 2107, 95 per cent were raised indoors, just 3.4 per cent were free range and 1 per cent organic. (The words 'free range' and 'organic', despite their wholesome implications, may guarantee less kindly treatment than consumers might hope for.) Ninety per cent of pigs were raised in intensive units with slatted floors and farrowing crates, metal devices designed to confine pigs during gestation and birth. Considered by many to be the cruellest of all, the dairy industry exploits cows to the maximum, increasing their annual lactations to increase milk yields, confining and artificially inseminating them, removing their calves too early, ending their lives at around five years old in the unremitting commodification of others' lives.

Resistance to improving these practices is invariably based on financial considerations. The share price of one supermarket group rose when at their AGM shareholders rejected a motion proposing that only free-range chickens should be sold. Recently, I heard a pig farmer saying that he is obliged to adopt lower welfare standards in order to meet the prices offered by the discount supermarkets. Discussions of international trade policies centre on the apparent imperatives for low food prices. At the same time, the huge enterprises behind the swift expansion in the numbers of CAFOs are extending their interests to the production of vegan products, just in case, and by doing so will ensure that it becomes even more difficult to escape their domination of what most of us eat. Meanwhile, CAFOs continue to do widespread and possibly irreversible damage to the environment in the spread of antibiotic resistance through their routine industrial use, in methane and nitrous oxide emissions and in the creation of putrid lakes, ponds and lagoons of molten excrement of the kind Jeff Tietz described so well and so revoltingly in 'Boss Hog – the Rise of Industrial Swine' published in *The CAFO Reader* in 2010. His descriptions recall the words of Lynn T. White: 'With ... the geological deposits of sewage and garbage, surely no creature other than man has ever managed to foul its nest in such short order.' Run-off from the industrial-scale raising of animals for food has created 'dead-zones', heavily polluted, toxic and lifeless, in the Gulf of Mexico, the Gulf of Oman, in Chinese lakes.

In *Animal Liberation*, published first in 1975, Peter Singer wrote of this method of food production as 'morally wrong'. In an article in the

Guardian in 2015, Yuval Harari described industrial farming as 'One of the worst crimes in history'. In his book *Eating Animals*, Jonathan Safran Foer critically investigates and describes methods of modern 'farming' and invites the reader to enter into the life of the creature inside a 'battery cage', one of a towering stack of similar cages three to eighteen tiers high, to imagine the sensation of being confined in an elevator so crowded that you're held aloft, above the foot-lacerating wire floor, in a place from which you will escape only when 'the doors will open, at the end of your life, for your journey to the only place worse'. It is a journey similar to those most of the birds and animals bred now for food will endure, ones whose lives are of subjugation to a system that's well known, witnessed, documented in its routine physical and mental assaults, in its casual violence, in the brevity of their existence, itself a crime against the natural span of a creature's life (but possibly, given the nature of that life, an amelioration too) in the Herodian, wanton mass destruction and discarding of male calves and chicks.

Even with growing awareness of the processes of meat production, and growing interest in veganism and vegetarianism, meat is increasingly produced industrially and still bought and consumed in enormous quantities. In a study of the earth's biomass distribution published in 2018, researchers determined that 'humans and livestock outweigh all vertebrates combined' – farmed poultry constituting 70 per cent of all the birds on earth, cattle and pigs making up 60 per cent of mammals.

Apart from other considerations about the effects on the well-being of the planet, whether we care or not may depend on what we believe about the abilities, feelings and perceptions of other species. The neuroscientist Lori Marino has written of recent research showing high levels of cognitive ability in chickens who, she concludes, 'possess a number of visual and spatial capacities ... some understanding of numerosity ... can demonstrate self-control and self-assessment ...

communicate in complex ways, have the ability to reason, perceive time, may be able to anticipate, are behaviourally sophisticated, have complex negative and positive emotions, evidence of empathy and distinct personalities.' Her research with pigs concludes that they too possess 'complex ethological traits similar but not identical, to dogs and chimpanzees'.

What about other creatures we eat? Fish? Squid? Octopus? Lobster? All of them are now known to be capable of feeling and perceiving more than we once could ever have imagined. The primatologist Frans de Waal has written of fish, describing himself as a 'lifelong fish fan' who keeps fish for whom he clearly has great affection. Using his own observations and research findings, he concludes that fish and crustacea feel pain, have emotions, can be frightened, depressed or bored. Does the consideration of feelings make a difference to human treatment of other species? Is it worse to kill and eat something we deem to be clever, capable of love and fear?

It stalks the mind, wakes the habitual worrier in the night. How can anyone do this? Where is empathy for other living beings? Is there meaning or sincerity in statements of concern about 'animal welfare'? Or is the word 'welfare' a sop, a cover, simply an expedient way to divert concern? How can we have moved the boundaries of our compassion so far? Where is an awareness in all this that our fate, and that of other species, is related to the totality of life on earth?

Often, it seems as if Descartes's 'beast machine', and the belief that animals are no more than insensate automata, shifted the sphere of the non-human with regard to consideration of sensation or suffering. Even though many of his ideas were founded on erroneous anatomical observations, and there was some misinterpretation of his writing, his influence became widespread although not universally accepted. A letter written to Descartes by the English philosopher Henry More in 1648 proved prescient in More's description of Descartes's method as 'deadly and murderous'. The philosopher and theologian Louise

Hickman believes that the ethical consequences of Descartes's ideas on other species were profound: 'The human relationship with nature is changed because nature is no longer to be contemplated for its beauty and wonder but is a machine to be used to human advantage.' In one of the most telling of his observations for *Parting Shots from Animals*, John Berger said: 'You used to sacrifice and kill us publicly. Today, your slaughterhouses are hidden away. It's not from us that you hide them is it? It's from yourselves.'

Once, even if we didn't actually see the slaughter, we saw the results. Shops displayed still recognizable body parts of cattle and pigs, and whole chickens, feet, head and all in their windows. Tracing back through my own life, it seems to have been the increase in supermarkets with their unique and sanitized ways of selling when the true alienation began, the distance between ourselves and our sources of food, our meat and the slaughterman, between us and the reality of an animal, once alive. I remember as a child, peering in revulsion at basins of tripe, now disappeared from butchers' windows, at the unidentifiable folds of strange, cream-coloured substance, gone along with tile floors awash with sawdust and the unmistakable, iron smell of blood. With that particular, sharp recall of scent, I'm back in the kosher butcher's shop in the Gorbals where my mother would drive to collect a parcel of bleeding brown paper containing meat prepared according to the strict rules of religious authority. In memory, I'm still dodging the meat alleys in the lanes of the Old City when as a student I'd walk home to East Jerusalem, trying not to look down the duskily lit stalls of blood-soaked limbs and heads and eyes, still trying not to see a young camel I saw being slaughtered once in northern Syria, avoiding that flash of the sight of bluish membrane and dusty fur.

The temples and altars of the ancient world, the Great Temple of Jerusalem among them, those bloody way stations in the narrative of how humans have killed animals, were destroyed or superseded by time, part of the development in how we've come to industrialize the lives and deaths of the animals we eat. The places where we killed them changed as we and our ways of living altered from those of opportunistic hunter to herdsman, from small farmer to industrial producer. The sites themselves have always mirrored economics and society, from the wild emptiness of Lindenmeier and its slaughtered *Bison antiquus* to the religious sites of many cultures, from farmyards and marketplaces to the modern, 'efficient' slaughterhouses of today. Nineteenth-century industrialization, with its attendant movement of population from country to city and the expansion of railways, altered patterns of consumption, and the preparation and slaughter of animals for food.

In America, the mechanization of animal slaughter began in mid-nineteenth-century Cincinnati in response to the swiftly expanding population catered to by a growing pig industry. River traffic was soon supplanted as it was overtaken by the growth of rail and the expansion of Chicago whose famous – or notorious – stockyards inspired Upton Sinclair's 1906 novel *The Jungle*, a lengthy account of the hardships and tragedies of a Lithuanian immigrant stockyard worker, his fellow workers and the unhappy creatures they dealt with. Although the book's main purpose was to be a rousing call to socialist action by detailing the horrors of the industry, the appeal to readers for sympathy for the workers became eclipsed by their sympathy for the animals. 'I aimed at the public's heart and by accident, I hit it in the stomach,' Sinclair declared. In a lecture in 1913, the English writer, vegetarian and social reformer Annie Besant described Chicago as 'Pre-eminently a slaughtering city' and was clearly aghast at the scale of mechanized animal killing.

Slaughterhouses were often sited near, or in the centres of cities, as visible, busy, thriving parts of the life and commerce of Chicago, Paris, Rome, Berlin, New York, Edinburgh and others. In 'Recollecting the Slaughterhouse', Dorothee Brantz describes contradictory attitudes towards the eating of meat and the means by which it was obtained, the high value placed on the eating of meat itself set against the contempt with which the slaughterhouse and its operators were regarded: 'Slaughter reflects the course of civilization, both in its continuity and change,' she writes.

One major development occurred when the city-centre animal slaughter of the eighteenth century and before, that noisy, noisome part of city life, began to be institutionalized and controlled during the years of the nineteenth century. The growth of urban populations in Europe and America had transformed landscapes, both rural and urban as the new railways brought people and goods into the hearts of cities, making the herding and slaughter of large numbers of animals insupportable. In a letter of outraged complaint to the *New York Times* in December 1865, one resident described the horrors of animal slaughter being carried out on Second Avenue: 'The blood and slime that flows upon this walk from the fountainhead of corruption . . . is such as to render the traveling most precarious . . . numerous degraded men and even women are ordinarily to be seen about the doorways of these shambles buying and haggling over the viscera of slaughtered animals . . .'

As slaughterhouses were regulated and moved from city centres, the herds of cattle, those ghostly assemblages ambling through streets in the haziness of old photographs, disappeared as the killing and all its horrible adjuncts, the fat rendering and glue-making, the skinning and flaying took place out of sight and scent, distancing creature from human still further, a process that was exacerbated by increasing urbanization. William Cronon's masterly book *Nature's Metropolis: Chicago and the Great West* details the history of the city's vast meatpacking industry and the effect of the trade during the nineteenth and early twentieth centuries in breaking the connection between eating and 'the moral act of killing', the relationship between man and animal: 'severed from the form in which it lived, severed from the act that had killed it, it vanished from human memory as one of nature's creatures'. The relationship was further becoming one of commodification with the animal becoming established as a potential source of vast industrial wealth.

Concerned too with questions of the relationship between money, power and how we treat others was the Spanish poet Federico García Lorca. While studying at Columbia University in 1929 – the year of the Great Crash – he wrote a collection called *Poet in New York*. The poems were observations of the life of the city, often fierce denunciations of cruelty, social injustice and the depredations of consumerism. In 'New York, Office and Denunciation' he wrote of the millions of creatures slaughtered in the city every day, the cows, pigs, ducks and chickens, their killing just another part of the onrush of capitalism.

In his paper 'Civilising Slaughter: The Development of the British Public Abattoir, 1850–1910', Chris Otter wrote of the totality of the split between the places where meat was produced and the 'polite world of sale and consumption'. An Eugène Atget photo in an old cookery book shows a poultry and game stall in Les Halles market in Paris in 1900; a stern-looking woman stands surrounded by furred and feathered creatures on hooks and slabs, necks dangling, unmistakably, distinctly what they are, what they once were. Recently, a supermarket chain announced that it is to introduce a wrapping that will protect the hands of the person about to cook a piece of meat from the apparently disgusting and possibly germ-spreading necessity of having to touch it.

As slaughterhouses were established further and further from city centres, the old sites – often large buildings, unaccommodating to any other enterprise and forbidding – were abandoned. Years ago, walking past the old Duke Street abattoir in Glasgow, by then long closed down, I inhaled an atmosphere of old suffering which seeped from it as it seems to seep from all such sites, from empty prisons, abattoirs, abandoned hospitals, spreading the legacy of melancholy and hazard, reminders of the dangers and inevitabilities of the physical and metaphysical worlds, as if they impregnate the walls, floors and corridors and in their dereliction empty windows are open again to suffering and death.

Writing of the redevelopment of Parisian slaughterhouses, Dorothee Brantz describes the building of the new site at La Villette in the north-east of the city in the early 1860s, of the elegance of the design of iron and glass, the functionality of the neoclassical architecture. The slaughterhouse at La Villette remained more or less unchanged throughout the twentieth century, closing only in 1974, later to become a centre for culture and the arts.

A long time ago, I watched Georges Franju's film *Le Sang des Bêtes*, a short documentary released in 1949, filmed in black and white at the La Villette abattoir. A careful, quiet work of art, it's also one of the most hauntingly shocking things I've ever seen, far more so than any of the films designed specifically to elucidate and persuade on the nature of animal slaughter. The abattoir itself is placed in the context of its peaceful, banal surroundings, the life of a quiet suburb, schools, churches, people, unfolding while the work of the slaughterhouse takes place. A beautiful white horse is shot without warning and swiftly dismembered, cows and calves are put to the knife, a line of headless lambs performs a macabre dance amid the horror. Steam rises from fresh blood as church bells ring and lovers kiss in the near but distant world outside while the impassive slaughtermen – in what seems like a symbolic withdrawal from the engagement of morality or guilt in the oppositions of innocence and violence – go about their killing. Like many other cultural figures, film-makers, philosophers and writers of his generation, Franju's work was deeply influenced by the events of the Second World War and *Le Sang des Bêtes* was made only a few years after its end.

Throughout a similarly quiet, low-key, brilliant work of art, shadows of *Le Sang des Bêtes* flicker. The book *The Search Warrant*, by the Nobel prize-winning Patrick Modiano, is the story of a young Jewish girl who disappears into the terrible streets of Paris in 1942, interwoven with the events of the author's own life in the city in the years after the war. In its reconstruction of the possible fate of fifteen-year-old Dora Bruder and other young Jewish women pursued by the French police and Gestapo, the book raises memories of betrayed innocence, of transportation and slaughter, of the extent of human cruelty and the consequences of indifference.

And now, in cities everywhere, the places where animals were once killed have been moved out of sight. 'The post-industrial age witnessed the demise of the modern mass-slaughterhouse,' writes Dorothee Brantz, 'because it did not fit into the image of the so-called post-modern city.' Madrid's Matadero, Rome's Testaccio, Les Halles, La Villette itself and many other buildings have been turned into centres for art, into cinemas and libraries, university departments, restaurant quarters and shopping complexes, the memories of their previous purpose kept alive in old signs and names, in the bull-wrestling angel statue of the Testaccio, the 'naves' of the Matadero. Does the sense of what once happened there lend the experience – of dining or strolling in a gallery, studying or drinking in a bar – a sharpened edge? Does a sense of past events remind us to live frantically, to appreciate, to blaze towards the unknown future? Does it allow the refinement of our sensibilities to further separate ourselves from the difficult world upon which we depend?

Once, decisions about how we obtain our food may have been less complex, determined by fewer considerations, more limited possibilities. People may have been moved or impressed by the idea of souls or suffering or the pain of others or may never have thought of any of them at all. It may then, as now, have been concern for health, adherence to a religion, a cult, a trend, a fad, a movement, the *zeitgeist* or something in the air that impelled people to follow a given course. It may have been the challenges of famine or poverty or the strictures of religion but whatever it was, it's rarely been simple. If, for many in the developed world in the post-war years, food hasn't been something about which much thought has been required, if abundance has made us careless and the commercialization of the food industry has provided us with ever more products to consume, our growing concerns about who and what we eat have become ever more real and urgent.

As if anticipating John Berger, in an article called 'Abattoirs', which appeared in the French intellectual magazine *Documents*, published between 1929 and 1930, the controversial writer and critic Georges Bataille wrote:

> The slaughterhouse relates to religion in the sense that the temples of the past . . . had two purposes, standing simultaneously for prayer and for slaughter. Nowadays, the slaughterhouse is cursed . . . and the victims of this curse are neither the butchers nor the animals but those fine folk who have reached the point of not being able to stand their own unseemliness, an unseemliness corresponding in fact to a pathological need for cleanliness.

'Cleanliness' is an interesting word in this context but as a concept it's dangerous, fraught with its own opposites, an open invitation to judgement, or worse. It has a contemporary air when it's related to food and current attitudes towards how and what we eat. The pursuit of the 'pure', the 'clean', the 'natural' in eating has become unaccountably inseparable from ideas of personal virtue and morality, as if one set of ideas necessarily indicates or predetermines the other, as if detached from the circumstances of our lives, the very act of our deciding what to eat or not places us on a scale of self-determined morality, one which often appears to belie the compassion which may be at least one of the motivations for abjuring animal flesh but appears less evident in attitudes towards dissenting members of our own species. The matter might be solely a personal one if it weren't so uniquely part of wider questions of economics, politics, gender and the ways in which we engage with all other species. (When Bataille talked of 'cleanliness', it reflected his own dark preoccupations with the transgressive and the shocking and his enduring fascination with the relationship between the religious and the secular.) Separations and conflicts surrounding our use of animals for food have opened up like the earth's own riven faults, extending into questions of spurious 'progress' and the need for conservation, into the dominance of the male, the uses of violence and the determined immovability of our political and economic systems.

Suggestions of an association between meat-eating and violence are long-standing, from Porphyry's belief that violence towards animals leads to violence against other humans, to the vision in the work of the twentieth-century American poet Robinson Jeffers. In his deeply controversial collection *The Double Axe*, published in 1948, he included the poem 'Original Sin', which, while purporting to be the description – a horrifying one – of a scene in which a mammoth is roasted alive by early humans, is in fact a howl of rage against the perpetrators of the recently ended war and an expression of despair:

This is the human dawn. As for me, I would rather
Be a worm in a wild apple than a son of man.

The belief that meat-eating is a necessity, or indeed a right, particularly for men, is deeply embedded in Western thinking about diet. Seen as indivisible from 'manly' attributes and pursuits, a food more highly economically valued than any other and believed erroneously to be essential for the promotion of health and vigour, meat has long provided both symbol and validation for masculinity which is difficult to avoid in every facet of history, culture and language. In an exchange in her novel *Jane and Prudence*, first published in 1953, Barbara Pym, a writer acutely attuned to the finer details of sexual politics, gently but magnificently mocked the idea of the supremacy of meat consumption. Jane, the wife of a clergyman, is discussing meals with Mrs Mayhew, the owner of the cafe where Jane hopes she can send her husband for lunch to avoid the onerous necessity of having to provide it for him every day:

> 'And the clergy are always with us where meals are concerned,' sighed Jane.
> 'Of course, a man must have meat,' pronounced Mrs Mayhew.

In her work, the writer Carol J. Adams unflinchingly examines the question of the objectification of both women and animals in Western culture, drawing parallels in *The Sexual Politics of Meat* between male power and the violence towards women and animals implicit in ideas of consumption and cultural portrayals of both. In her book *Burger*, she analyses the language of the hamburger trade, drawing attention to the far from subtle implications of names such as 'Big Mac', 'Whopper', 'Beefy Boy' or 'Super Boy' and the other similarly crass ways in which meat is sold both as food and encouragement to male sexual vanity. A very upmarket New York deli offers Christmas gifts 'for guys': enormous sirloin steaks, huge, vastly expensive ribs of prime beef.

Elena Ferrante's redoubtable protagonist Lila in *Those Who Leave and Those Who Stay* is assaulted by her boss while working in a factory that manufactures meat products. It happens in the room where salamis are hung to dry and as he makes his ultimately unsuccessful attempt, he invokes the odours and appearance of the meat around them in graphic sexual terms, in a scene that perfectly illustrates the objectification of the powerless and the symbolic identification of women with meat.

The act of the slaughter of beasts itself is often seen as another aspect of manhood, a further expression of the qualities such acts are deemed to require – courage, strength, *machismo, cojones*. Whether by hunting or by the act of the slaughterman, killing's all too often seen as a validation of masculinity, creating an unending connection between killer and killed, sexual power and violence. In a recent essay, the Dutch writer Arnon Grunberg, working in slaughterhouses in the Netherlands and Germany to observe and document his thoughts, writes of standing in a pool of blood, considering if he'll break up with his girlfriend: 'Then I change my mind: I want to fuck her to death. I will turn the bed into a sea of entrails, lungs, kidneys, blood and shit. Never have I been as horny as here, in the slaughterhouse.'

Mostly, we require other people to kill and prepare the animals we're to eat but does it make any difference to our meat-eating if we do the killing or witness it ourselves? It is often suggested that watching animal slaughter is a necessary adjunct, or precursor, to eating meat, or a possible deterrent. But does it validate subsequent consumption? Might it be an instrument of desensitization? Does buying meat that has been prepared, packaged and date-stamped, render the buyer beyond an ethical stance of someone else's arbitrary choosing? What difference might our knowledge make at the point of our buying meat to consume? I'm unsure if viewing animal slaughter matters, if it reduces the supposed hypocrisy, changes the fundamental nature of the transaction. Watching killing may indicate one's own level of supposed bravery or bravado, or may validate or encourage a tendency towards voyeurism. It may allow an understanding of the process and the experiencing of particular sensations but moral righteousness is of a different order entirely. It's difficult to believe that obliging oneself to witness the events of a slaughterhouse in any way reduces or increases a possible tendency towards hypocrisy.

Plenty of accounts have been given, filmed or written by people who have worked in slaughterhouses or spent time there as observers. Some, like Grunberg seem unmoved, or moved in strange ways (although to be fair to him, he does decry the poor conditions and pay which are universal in the industry) while others are upset and disquieted. In her book *The Ethical Carnivore*, Louise Gray documented the year she spent eating only meat she had killed or had seen killed herself, describing her feelings after visiting a slaughterhouse to see the processes involved and the reaction of one of the slaughtermen who suggested 'not without satisfaction' that the experience had traumatized her, which it seemed it had. Throughout her accounts, there's an underlying sense from those who work in the trade, that a lack of characteristics they view as predominantly male – physical strength, detachment, unsentimentality – constitutes some kind of regrettable weakness, as if it's a test, as if there are those who pass and those who fail.

I would, without question, fail. *Le Sang des Bêtes* stays with me still, as does the belief that people suffer from witnessing suffering, that it damages us or inures us to brutality, that seeing too much ruins the imaginative capacity, that without horror being new, it's old. Sights stay and haunt in small ways or large. Having been forced by a teacher to visit a slaughterhouse, the Nobel Prize-winning Austrian writer Elias Canetti, a man always concerned with the treatment of other species, suggested that watching the killing of an animal was not something he was 'meant to get over'. Viewing the suffering of the small animal, the bird, the human, make us the same in our extending whatever feelings we have. An image from 'That Rat's Death' by the American poet Eileen Myles reminds me of the recurring, indelible nature of witnessing violence as she writes of seeing the rat's squashed remains over and over in her mind in the days after she witnesses his death.

Questions stay with me – what can be inferred about us from what we choose to eat? What does it say about a person if they're carnivore, vegan or vegetarian? Does what we choose necessarily indicate anything beyond our choice of food? Do veganism or vegetarianism necessarily suggest kindly munificence, tolerance, care for the environment and for others, human or not? What about political affinities? A traditional association exists between vegetarianism and progressive thinking – newly released from imprisonment for suffragette activities, the heroine of H. G. Wells's novel *Anne Veronica* is borne in triumph to the 'Vindicator Vegetarian Restaurant'. The vegetarianism of George Bernard Shaw has often been seen as synonymous with the politics of the time, with a new socialism and concern for the rights of others but is the view correct? Do vegetarianism and veganism necessarily indicate anything about our propensities for virtue? If they do, which and what and how? They may, but then again, they may not.

Among my own random, too infrequently edited collection of cookery books, is the one in which I recently came across this passage:

> Meat (the flesh of slaughtered animals) is ... poor in vitamins. Some vitamins are stored in certain organs of animals e.g. in the liver, but these organs are for other reasons unfit food for man. All kinds of meat as well as fish and poultry, bring about a slow decay of the vital tissues of the human organism. Animal flesh nourishes beasts of prey for they devour it living, with the blood and bones but butcher's meat, cooked and prepared for the use of man, is no proper food for him.

This definitive if arguable statement is from the book *Health Giving Dishes* by Bertha Brupbacher-Bircher, manageress of her husband Dr Maximilian Bircher-Benner's sanatorium in Zurich. (Famously, he was the inventor of 'Bircher müesli'.) The book, an exposition of the doctor's ideas and published in 1934, belonged to my grandmother, an apparently enthusiastic follower of fashionable trends in dietary thinking of the time. The recipes in the book are vegetarian, punishingly plain and the tone of the writing is brisk, probably a reliable indication of the nature of the sanatorium's regime. After he'd stayed there and been obliged to take part in the *Ordnungstherapie* of early morning walks and cold-water treatments, the writer Thomas Mann described the clinic as 'a hygienic prison' (and possibly used it as inspiration for the institution in which he subsequently installs his character Hans Castorp for an extended stay in *The Magic Mountain*).

Dr Bircher-Benner began his sanatorium under the strong influence of the 'Lebensreform' movement which had begun in Switzerland and Germany in the late nineteenth century in response to the socially and economically restrictive effects of growing industrialized urbanization and was manifested variously through diet, new farming methods, exercise and the enthusiastic if sometimes frankly odd embrace of 'nature'. Spreading rapidly across Europe and the United States, Lebensreform became popular and lasting, its ideas still evident in many current attitudes towards food, health and nature. It was unfortunate although unsurprising that aspects of its ethos became mingled with the social thinking of the German nineteenth-century romantic nationalist *Naturephilosphen* movement to become at least partially absorbed during the twentieth century into the odious ideologies of 'racial hygiene', the Blood and Soil movement, the Artaman League and ultra-nationalist land ethic promoted by Walther Darré, Reich Minister for Food and Agriculture (one of the most prolific kleptocrats of a vastly kleptocratic regime) and the sinister Nazi admixtures of the bucolic, the narcissistic and the psychopathic. The apparently innocent pursuits of naturism, 'eurythmy' dancing and hiking were overtaken by the Hitler Youth movement, the fierce body-perfectionism of the Nacktkultur movement and growing elite military culture.

The biodynamic and organic horticulture of the Lebensreform which encouraged vegetarianism, soil improvement and small-scale agriculture, became a topic of interest not only to Darré but to Hitler's deputy Rudolf Hess and Heinrich Himmler, influencing government policy towards forests, agriculture, the treatment of animals and land use. During the Third Reich, biodynamic gardening was encouraged (even at Dachau and other concentration camps) and, with almost inconceivable irony, strict animal protection laws enforced. The reasons are complex and are explained in the paper 'Understanding Nazi Animal Protection and the Holocaust' by Arnold Arluke and Boria Sax as having their roots in the personal attachments many leading

Nazis had for their pets, the identification of Germans and animals as similarly oppressed and threatened, and in an altered view of the demarcations between humans and animals, which allowed the contradictory attitudes involved in carrying out extreme cruelty towards humans while extending protection to animals. Hitler certainly abjured eating meat for some time at least, believing its physical consequences to be the roots of decadence. Some reported having observed him eating meat while others suggested that he limited his meat consumption in an attempt to cure the flatulence which, according to a social history of the period, was an 'unpleasantly noticeable phenomenon in public places' caused by the eating of wartime bran-bread.

In Britain at the same time, an ideological nationalist land ethic was being promoted by a small group of oddities, among them Henry Williamson, writer of *Tarka the Otter*, and member of the British Union of Fascists, sundry ultra-right wing back-to-the-land-ists and like-minded supporters of Sir Oswald Mosley (including the infamous William Joyce, later hanged for treason). They idealized a type of anti-urban, volk-isch feudalism and were, I presume, prevented from becoming mass murderers as many of their counterparts in other parts of Europe did, by virtue of geography and lack of opportunity. While many of their ideas on agriculture had a great deal to recommend them, their political ideas were generally undifferentiated from their repellent cohort in Germany and while as individuals they faded from view after the war, their influences and ideas can still be discerned in various organizations they helped found connected with ecology and farming. In the excoriation of the urban and the expression of an atavistic, sometimes misogynistic and exclusive nativism, echoes of their ethos still sound in at least some current thinking in environmentalism and writing about the natural world.

Concerned about ethical issues involved in the choice of meals served during conferences related to the field of animal studies, one academic recently questioned the participant's requirement to 'confess' their dietary choices, vegan or otherwise. With her unease over the encouragement of oppositional views of the sort which place you irredeemably on one side or another, Traci Warkentin expressed in an essay a wider disquiet over the nature of conversation and debate in the postmodern world, what she referred to as 'reversed dualism' in the positioning of people as either 'for' or 'against'. This strand of thought is exemplified in a current book of philosophical essays on the ethics of food, which earnestly promotes veganism as the sole moral ideal. One contributor concedes that there are difficulties to be faced by those who wish to be vegan but cannot afford the foodstuffs required for healthy vegan living. She acknowledges that for such people, much greater expenditure of time, energy and resources would be required to attain the moral ideal more easily available to the wealthier. Her suggestion for dealing with this deficit would be to give the disadvantaged who seek to maintain a vegan way of life 'moral passes' to make up for falling short on their moral duty. The only problem she appears to see with her proposal is that such people might not like being considered 'morally subpar' and might feel a sense of guilt and inferiority in the face of the moral superiority of the wealthier, 'true' vegan.

In 2004, the late Val Plumwood, the Australian feminist environmental philosopher, wrote a paper in which she discussed two opposing eco-feminist theories on the human relationship to other species and the environment. Always critical of the division in Western society between man and nature, body and mind, she was deeply concerned with the effect this dualism had on both the natural world and on the marginalized. Acknowledging the opposition on both sides of the debate to the modern system of factory farming and abuse of animals, she examined the concepts of 'ontological veganism', which advocates 'individual consumer abstention from all use of animals' as a means of protecting animals from factory farming and other harms, and 'ecological animalism' which recognizes humans as part of a larger context of environmental and social awareness. In her critique of the first, she suggested that it encourages a 'human/nature dualism', effectively removing humans from the world of other species. She described the position as 'ethnocentric', 'outside nature', centred in Western urban thinking, which favours a 'privileged "consumer" perspective' while ignoring or marginalizing other cultures, peoples and social groups.

'Ecological animalism' advocates a semi-vegetarian diet, drastic reductions in first-world consumption of meat and opposes the kind of treatment to which animals are subjected in an industrial farming system. This way of thinking, Plumwood suggested, offers a perspective which avoids the gender and social stereotyping that excludes the lives of the poor and disadvantaged or indigenous communities, where often women play a part in activities such as hunting, while regarding humans as only one part of a larger ecological system. 'Ecological animalism', she wrote, 'insists that we must consider context to express care for animals and ecology, and to acknowledge at the same time different cultures and individuals in different ecological contexts, differing nutritional situations and needs.' A woman of extraordinary courage, Plumwood was attacked by a crocodile while on a kayaking trip, managing to crawl for miles to get help. Writing of the encounter, Plumwood described contemplating the fact that she was on the brink of becoming nothing more than another creature's food, a realization that made her believe that every creature is 'more than just edible'.

I read the responses to a vacuous, airy interview with a 'celebrity' on the subject of becoming vegan. When he declares 'It's easy to give up animal products' people respond on social media by telling him why it's not. Their stories are humbling, of poverty, gnawing worry and need, of eating cheap burgers and chicken from necessity, of being able to survive only with the sustenance bought from outlets the wealthier have the funds to be able to despise. The 'degraded' poor, scrambling and fighting over slimy viscera in a New York street in 1865, one can safely assume, did not do so through choice.

In *Eating Animals*, Jonathan Safran Foer writes of two friends ordering lunch. One decides on a burger because he's in the mood for one. The other, although he's in the mood for one too, desists because he remembers 'that there are things more important to him than what he is in the mood for at any given moment . . .' and thus, Safran Foer presents the dilemma central to every debate on what we eat and what we don't. What *are* the things that are important to this particular non-burger eater? Self-restraint? Concern for the environment? For non-human creatures? Which ones? The commendable desire not to contribute even the cost of a burger to a system that keeps some people poor while allowing others the bounty of free choice? But 'being in the mood' for anything at any given moment, along with the opportunity to indulge, or repress the mood, is a gift of fortune that perhaps not everybody has.

The implications of ascribing moral force to food choices extend beyond questions only of food. In 1968, the anthropologist Mary Douglas wrote a book called *Purity and Danger*, now a classic of anthropological literature. In it, she discussed concepts of taboos, cleanliness, defilement and the boundaries individuals and societies set for themselves and others in matters relating to every aspect of human life. In her introduction to a revised 2002 edition, she writes of the political climate that surrounded the first publication of the book, a time she said when formality was being rejected and various types of subjugation – sexual, racial and colonial, were under scrutiny; not the best time, she says, for the publication of a book that discussed matters of constraint and control.

By the 1970s, the 'passionate moral principles' of the 1960s had transmuted into the fear of contamination: of air, water, food, leading to questions of risk, danger and the nature of power. Reading her book makes me wonder about current responses to food culture in all its forms, if they're symptomatic of our times, ways of confronting and trying to gain control of the uncertain present, as well as a means to express many sorts of anger. It may be an attempt to counter the threats to the planet and its species, including ourselves, in whatever way we can, one definitive measure we feel we can take in the face of the vast implications of the uncertain, changing world. The dangers are ones I'm reminded of when I look at the picture of the killing knives that intrigued me that morning in the Jewish Museum.

The leaflet from the blood exhibition lies on my desk. 'Uniting and Dividing' it says. The photo of those knives, used in the slaughter of animals for food, faces me with another set of divisions and the question about the eating of meat, which has long surrounded *shechita*, the method by which in Judaism animals are killed to conform with the strict laws of Leviticus. Often wrongly called 'ritual slaughter' (the word 'ritual' hinting that the slaughter itself might be part of some kind of strange, antique and possibly sinister ceremony, which it's not), a more correct term is 'religious slaughter'. Similar in many respects to the traditions of Islam and *dhabihah*, the method of slaughter by which halal meat, allowable for Muslims, may be obtained, the foundations of both lie in ancient Jewish and Islamic prescriptions of attention to animal welfare, representative of the ways in which our lives and traditions coexisted and overlapped in the countries where we lived together for so long.

Controversies surrounding both kosher and halal meat are based on these methods of slaughter and whether or not they're cruel, or more so than the ordinary, day-to-day methods, the captive bolts and gassing, the electrocution and drowning. The difference in the method of slaughter is that animals are not stunned (although certain stunning methods are allowed under Islamic law) and their throats cut according to very precise laws determining how this should be done, by whom and with what, with the intention of minimizing suffering to the animal. The swift killing is the first concern, the equally swift dissipation of blood is the second. 'Be resolute in not eating blood,' 'blood is the soul' it states in Deuteronomy. To be properly kosher, meat must, in addition to being slaughtered by *shechita*, be salted and steeped and washed several times to rid it of all blood.

For Jews, blood is the ultimate in taboos – it's forbidden to eat the blood spot in an egg and there is considerable rabbinic discussion about whether or not one is allowed to suck a cut in one's own finger. (To be found on religious question-and-answer pages under 'Is my own blood kosher?') All of this makes it seem like the oddest of anomalies, as well as the most profound of calumnies, that the very substance chosen by the many illiterate fantasists, sundry vengeful psychopaths and violent, ignorant racists who have appeared over centuries to try to implicate Jews in the commission of wanton and egregious crimes, has been blood in the phenomenon known as 'blood accusations' or 'blood libels' – sadly not yet a relic of history, which suggests that Jews, requiring the blood of Christians for unspecified, sanguinary rituals, acquire it by means of murder, often of children. The conflation of 'blood libels' and the idea of 'ritual' slaughter and supposed cruelty is still common, appearing in a new form, directed this time and with similar ill-intent against Islam and its practices.

None of the questions about the killing of animals for food by any method can be separated from the industrialization of food production as a whole. The conditions laid down in both Judaism and Islam imposed rules before anyone knew what they do now about circulatory systems and blood and pain and the relationship between them. It was, and is for both traditions, a profound belief that humane treatment was a benefit, that preventing those deemed unsuitable for the task from being responsible for the slaughter of beasts, and that a very sharp knife used by an educated and devout person, were all appropriate measures to ensure that the laws on animal welfare could be obeyed.

I'm not sure how I can judge whether the intention of either type of religious slaughter, designed to minimize suffering to the animal, is borne out by facts, ones difficult to come by in any usefully objective way but the discussion has come to represent something far beyond questions of kindness to other creatures, human or not, which should be the requisite of moral and decent humans everywhere, more than just one of how the religiously observant might obtain the only meat permitted for them to eat. The evidence is often as partisan as the arguments and the question of religious slaughter rarely very far from political questions of rights and freedoms, which have, tediously and for far too long, provided a ready excuse for the denial of both as well as a precursor and warning sign for communities long-attuned to the results of illogical, ill-intentioned conflations of religion and cruelty and to the possible aims of those who try to limit expressions of the religious freedoms, and sometimes even the existence, of minorities.

An important question is how animals are treated during their lives and the processes of their deaths, and since animals destined for both kosher and halal consumption are the same ones destined for the general market, the boundaries of the questions expand and change. Incidents of animal cruelty beyond even the baseline cruelties have been reported during both methods of slaughter. Given that the majority of methods of modern-day, industrialized animal raising may violate the laws of *tsa'ar ba'alei chayim*, 'the suffering of animals', and some aspects of Islamic law, even before the animal reaches the point of slaughter, if one is to be strict in the observation of laws as far as one's own practice is concerned, thought might be extended to the question of how the animal is reared.

Years ago at a book festival, I spent time chatting to the members of a Jewish vegetarian organization. We discussed animals and birds and how Jewish thinking and scripture might predispose us, via the earlier versions of Genesis, to vegetarianism and as I usually do I probably talked at length about the intelligence of birds (particularly my own) and although I can't remember now, we may have talked about examples from the past and some of the great figures of the religion who were known vegetarians. If we didn't, we missed an opportunity.

Many rabbinical scholars have written and commented on ethics and our relationship with animals and food, from the writers of Leviticus onwards and many have advocated vegetarianism as the most appropriate way for Jews to live. Among the most intriguing are those who followed the religious ideas and way of life of Kabbalah, the Jewish mystical tradition, that mind-alteringly complex, esoteric, profound, mad and often lovely endeavour to understand and draw closer to the source of creation. In its attempt to transcend the material and the self in the attainment of enlightenment, it's been likened in some of its ideas and practices to Taoism, the calm, ever-chilled path upon which Lao Tzu has, for 2,500 years, tried to lead us all to an understanding of how to live best in the universe.

Possibly the most ardent of Kabbalist practitioners belonged to a group who, during the sixteenth century, gathered from all over the Mediterranean world to settle together in the Galilean hill town of Safed to live and study and practice the ways of Kabbalah, among them the most learned of scholars, rabbis Isaac Luria, Joseph Caro, Solomon Alkabez, Moses Cordovero and others. These men and women embraced ideas of compassion and the one-ness of all life on earth, modelling their lives on that of the Essenes – famously ascetic, communitarian and vegetarian – working, studying and tending their gardens.

It is an ideal, unattainable and distant, altered by history and a knowledge of the exigencies of the current world but still I hold it in my mind. I remember a passage in Perle Epstein's book on Kabbalah where she writes of those religious scholars preparing for their vegetarian Friday night and the *Shabbas*, dressed in white with leaves in their hair, dancing in the streets, watched by the many cats who, Epstein says, are still there in the small, beautiful town, regarded by some as the reincarnations of those dancing mystics, silent observers perhaps, of the swift passing of centuries or the unfathomable ways of *Homo sapiens*.

5

Rights

There's a haze today, sea mist and stillness. The smooth cobbles shine with an ancient glaze. Time dissolves, spreads into the cold darkness. Late afternoon and by now everything of the day has lulled and slowed. Here, the world's continuous fixation with numbers and schedules and the urgency of the moment seems to melt, centuries turning to breath in frozen air. The dark figures wandering past me through the narrow lanes between buildings, through cloisters, under ancient walls, have been here, just like this since the twelfth century, walking the same paths, on the same endless repeat, the same laughter, the same voices talking as they've always done, learning, seeking out what is or might be in the worlds of spirit, thought and substance. If the figures swaddled in coats and scarves against the cold are ghosts, they're ghosts with memory and habit, the old ones of drinking and learning, the obligations and sometimes delights of scholars and students everywhere and always. I like being here; I like the buildings, the stone, the ivy on the walls, the atmosphere of this ancient university catching at me, the inviolable sense that the very air instils, if not knowledge itself then, at the very least, a welcome tendency towards thinking.

I've come here to walk and linger a bit at the place where in the eighteenth century there was an inn called the Old Red Lion. At 5.00 p.m. on every second and fourth Wednesday of the month in the years between 1758 and 1775, some of the most distinguished scholars of the Scottish Enlightenment met here. They called themselves 'The Aberdeen Philosophical Society' or 'The Wise Club' and their aim was to investigate: 'every Principle of Science which may be deduced by Just and Lawfull inductions from the Phaenonomena either of the human Mind or of the material World ...' rhetoric, the sciences, medicine, mathematics, the slave trade, the natural world – including in 1771, a discussion on 'How are the Proceedings of Instinct to be distinguished from Reason or Sagacity in Animals?'.

It's too cold to hang about but I do, just long enough to imagine stopping them one by one as they go into that long-ago hostelry to clutch at their material or immaterial sleeves. I feel as if I'm waiting for some kindly, revenant scholar to stand me a drink and share their thoughts, have them tell me if they believe there might be something here in the air or stones to connect me with that other time and the discussions that took place here, one of these moments when you might hear ancient voices in the icy air.

Two books have brought me here. They've stood side by side on my bookshelf in uneasy symmetry for a long time and by now I can't remember if I bought them at the same time, and in the same second-hand bookshop. Both were published decades before I did. I didn't read them straight away – they're the kind of books you chance on and buy thinking they're bound to be useful later and are delighted to unearth when eventually they are. The symmetry is in the titles: *Man and Mouse: Animals in Medical Research*, first published in 1984, was written by Sir William Paton, widely regarded as one of the world's greatest pharmacologists. The other, from 1986, is *Animals and Man*, written by Dame Miriam Rothschild, zoologist, flea-expert, all-round scientist, vegetarian, humanitarian, defender of those who should be defended, remembered for the white Wellington boots she wore in lieu of leather shoes and writer of this brief, scholarly, witty lecture on the way we should treat animals in every sphere of life. The first book is a precise, detailed, sometimes scientific defence of experimentation on animals; the second, although written by someone who has taken part in it, is deeply critical of the methods, extent and justifications for its use. A lot has changed in the years since the books were written but not enough to alter the fundamental questions – how did we reach this point of using other creatures in the ways we do? Where are the boundaries of the things we may and may not do to others?

In one way, looking back on the subject is an exercise in that particular variety of despair anyone may feel on reading any aspect of history. An accumulation of questions, disquiet and often disbelief at what humans can do to one another and to other species leaves me feeling uncertain and unresolved. I know too that without the countless creatures, human and not, alive and dead, offered up to the will of knowledge or progress, we wouldn't know about bodies, disease processes, what's safe, beneficial, life-saving, or not. To ask whether it's justified to use animals for any purpose is just another demonstration of the way in which the foundation of our lives is built on our indivisible relationships with other species, another strand of exploitation, of the cruelty, love, desire for benefit and every other difficult, convoluted emotion and belief with which we live.

The years during which those Enlightenment scholars among whose ghosts I wander met to debate were significant. Views of humans in relation to other species were among the ideas that began to proliferate in the dazzling intellectual atmosphere of the late seventeenth and eighteenth centuries in Europe. Challenges to previous certainties of human supremacy, superiority and God-appointed place in a man-centred universe were being eroded by the expanding understanding of cosmology, the sciences, philosophy and a widening of the boundaries of the physical world. Philosophers and political thinkers were considering fundamental questions of how to behave towards others, one human to another, individuals in relation to the state, states in relation to one another, humans and their relations with the Divine and in their dealings with and treatment of other species. 'Nature' itself was undergoing a change from God-ordained to the subject of humankind's multifarious interests and attentions in representing the antithesis of the corruption of culture and society. Ideas of 'natural law' and 'natural rights' ascribed to humans were expanding to include the possibility of those of animals. The accepted Christian world view of Aquinas et al. was being questioned along with the views that denied feeling or reason to animals and gave humans the freedom to act towards other species in any way they chose.

The philosophers of the Scottish Enlightenment – judges, ministers, teachers, scientists, artists and artisans inspired by the atmosphere of the day – were among those who had begun to reflect on every aspect of human life, including the human relationship with the natural world. One of the most idiosyncratic, William Smellie, an Edinburgh printer, friend of Robert Burns and outstanding, self-taught natural philosopher, translated the Comte de Buffon's controversial *Histoire Naturelle*, incorporating some of it into his own extensive work, *The Philosophy of Natural History*, in which he suggested (possibly correctly) that the cooperative behaviour of beavers makes them better democrats than humans. Lord Monboddo, a prominent judge, made studies in anthropology, formulating theories on the nature of animal and man, suggesting that orang-utans, as a form of human, might be taught to speak. The distinguished academics and thinkers John Gregory, Thomas Reid and Adam Ferguson all considered questions of moral obligations towards other species while Frances Hutcheson has been credited, in questioning the ways in which humans used animals, with providing the first arguments in favour of animal rights. David Hume and Adam Smith, more famous now for their works on economics and politics, both wrote on the subject. 'No truth appears to me to be more evident, than that beasts are endow'd with thought and reason as well as men,' Hume wrote and although he expressed qualifications about the limited capacities and reliance on instinct of other species and the superiority of humankind, he wrote too of the mutual 'sympathy' which exists and is communicated from one animal to another and from animal to human.

Adam Smith, in his *The Theory of Moral Sentiments*, expounded ideas of the extension of sympathy although he was little different from others in extending his sympathy only so far. In *The Wealth of Nations*, animals are described in terms of value, referred to as 'unmanufactured commodities' or in the case of domestic animals as creatures which 'human industry can multiply in proportion to demand', a view of the natural world as a source of human economic prosperity which seems to presage, if not encourage, the exploitation of animals as 'resources' and the development of ideas such as those of 'natural capital'. Writing about the contribution made by the thinkers of the Scottish Enlightenment to changing attitudes towards animals, Nathaniel Wolloch suggests that the view of 'animals as a resource vital for the cultural progress of humanity ... did not bode well' for the future of animals in the modern world.

Trying to understand the nature of our physical being has always been fundamental to us, the ways in which our own and others' bodies grow and change and reproduce, blood and hearts and lungs meshing their unseen workings into our mysteriously functioning whole, how our brains command, or don't. We've always wanted to know what goes wrong and why, how to cure and heal. Our discoveries have been made by examination, observation, dissection, vivisection – the cutting-up of dead and living beings, human and animal, and for as long we've been doing it, there has been profound disquiet about how it's done, and why.

Fragments of papyrus from Egypt appear to indicate that at least some of the origins of human biomedical science lie in early Egyptian veterinary medicine and the attention paid to maintaining the health of the temple and sacrificial cattle. The Egyptologist Andrew H. Gordon believes that the significance of the role of animals and those who cared for them in the development of medicine, has been downplayed by researchers who historically had less regard for matters veterinary, referring to vets as 'butchers', rather than the expert medical practitioners they were. The information lies in the often crumbling, damp-stained, insect-nibbled archives, frequently named for the Victorian gentlemen who found – or bought them after someone else had found them – Kahun, Edwin Smith, Ebers, Hearst, or the Ramesseum papyri, written mainly in hieratic, the hieroglyphic-derived cursive script of the time, texts on matters medical, surgical, gynaecological, paediatric and veterinary. Many contain acute medical observations, diagnoses and treatments little different from today's. The oldest, the Kahun, may date from Egypt's Middle Kingdom, around 1900 BCE (although it is suspected that they're considerably older) and include treatises on human and veterinary gynaecology and medicine. In a recent edition of a journal of spine injuries, an article on the Edwin Smith Papyrus, which dates from 1600 BCE, suggests that the contents would be regarded as 'state of the art' clinical practice today. I ask David, my now ex-husband, a retired neurosurgeon, to read them and he agrees with their opinion. We both then comment sagely on a theme of 'nothing new under the sun'.

Interpreting the workings of the body by post-mortem dissection has never been straightforward. Until today, the treatment of corpses is surrounded by stigma and wariness, death being as it is suffused with myth and ritual, taboo and fear. Ancient cultures held death apart with awed and fearful respect, with rules, restrictions to avoid defilement and pollution, and strict views on the appropriate ways to deal with corpses in order to satisfy the social and spiritual imperatives of present and future. Connections between body and soul, possible requirements for a future post-death life and the relationship between the living and the dead, all affected the treatment of the body after death. In many early civilizations, human dissection was forbidden, although the desire for knowledge seems to have overcome at least some of the strictures. In Greece, societal taboos and religious objections prevented dissection of humans although dissection of animals was carried out, as was some limited investigation of human bodies in the clinical context of surgery. Alcmaeon of Croton, described as 'the Father of Anatomy', is credited with establishing the connection between the brain and sensory organs and identifying the function of the optic nerve, possibly through vivisection or dissection of animals, although argument surrounds the very fragmentary evidence available. Aristotle believed that because our own inner workings were obscure to us, we had to learn from those of other creatures. He was an extensive and thorough vivisector and dissector of other species, working alongside his wife Pythias, who had an interest in embryology and histology. (Infrequent mention of women in early science is not because women had no part in dissection and studies in anatomy or because they did not have distinguished careers in science and medicine, but because their work is, unsurprisingly, less known. In an extensive and fascinating volume published in 1938, the doctor and historian Kate Campbell Hurd-Mead documented the history of women in medicine from ancient times until the eighteenth century.)

Diocles of Carystus, Aristotle's probable contemporary, wrote prolifically on many topics concerned with anatomy, physiology, embryology, gynaecology and nutrition, studies most probably based on animal experimentation. Praxagoras, the discoverer of the difference between veins and arteries, also experimented on animals but failed to achieve the fame, or perhaps notoriety, of his pupil, Herophilus of Chalcedon. Regarded as one of the foremost figures in the study of anatomy, Herophilus, while working with his contemporary Erasistratus in the intellectually free and vibrant milieu of Ptolomaic Alexandria, is reputed to have made studies on live humans, possibly convicts, in what may have been the first systematic examination of the living human body. Whether or not these suggestions of human vivisection were true, the impact of the mere possibility sent horror and distaste through time – in the first century, the Roman Aulus Cornelius Celsus wrote in his medical encyclopaedia *De Medicina* of Herophilus's use of living humans as 'brutal and unnecessary'. One hundred years later, the Carthaginian Christian theologian Tertullian, greatly outraged by the thought, said much the same except that he called Herophilus 'a butcher', an intimation of the belief that if such things were done, they represented the nadir of human iniquity. Herophilus and Erasistratus's dissections, of whatever nature they were, were unique too in being both the first and last in a thousand years in the history of Greek science when humans were dissected, or vivisected. Cessation of the practices was brought about by the religious and social constraints that prevented such work being undertaken on humans either living or dead.

In the Roman Empire, dissection of human corpses appears not to have been undertaken, whether because of the law or the effects of social stigma, obliging one of those who became a notable figure in the study of anatomy, Aelius Galenius – Galen – to work only on animals. Born in Pergamon in Greece in 130 CE, Galen spent most of his life in Rome as a physician to gladiators and a writer on medicine and anatomy. A prolific experimenter, he worked on dogs, apes, pigs and cattle and became an enthusiastic demonstrator of public and competitive vivisection – including the reportedly hideous public evisceration of a monkey.

Human dissection may have taken place, albeit secretly, in the 1,700 years during which it was expressly forbidden by the dictates of society and later by the church, but there seems little record of it. In 1231, the remarkably forward-thinking Holy Roman Emperor Frederick II decreed that a human body should be dissected once every five years for the furtherance of knowledge of anatomy. This declaration was followed in several countries by the lifting of the ban in the later years of the thirteenth century and early years of the fourteenth.

The first public demonstration of dissection, heralding a return of the practice, took place at the University of Bologna in 1315 and was carried out by the professor of surgery Mondino de Luzzi, who influenced the extensive anatomical studies carried out by Leonardo da Vinci. Some of the beautiful anatomy theatres where these demonstrations were carried out still exist, including those at Bologna University and in the Palazzo del Bo at Padua University where one of the best known of anatomists, the Flemish Andreas Vesalius studied and worked. Vesalius, born in Brussels in 1514, was reputed to have begun vivisecting animals as a child – a slightly disconcerting thought – moving on to human dissection later. Disproving much of Galen's work, he wasn't restrained in commenting on Galen's errors as a result of only being able to dissect animals and is reputed to have said that Galen was 'fooled by monkeys'. Vesalius's contribution to knowledge of human anatomy and subsequent human medicine was vast, particularly in the fields of cardiology and cardiovascular surgery, possibly helped by his willingness to benefit from clandestine gallows-snatching. (The shortage of a ready, legal supply of suitable bodies has always proved a practical and moral difficulty for the would-be dissector.) Vesalius was a flamboyant public demonstrator of the arts of dissection and vivisection as well as an accomplished self-promoter. His magnificent book *De Humani Corporis Fabrica* was published in 1543. Monumental and beautifully illustrated – as well it might have been since some of the three hundred or so illustrations are thought to be the work of Titian or one of his pupils – it contained, modestly enough, a frontispiece in which Vesalius had himself portrayed in the distinguished company of Aristotle, Galen and Hippocrates.

Public dissection became popular from the sixteenth century onwards, expositions abounding in words as well as in performance, in depictions of the watchers watched, caught in what might have been their avid desire for knowledge or their voyeuristic fascination. If attending public dissections seems a strange way to spend time, public vivisection seems entirely beyond that, as inexplicable (to me anyway) as any of the extravagant spectacles of the killing of animals in early societies or the public executions, which have taken place until today. Perhaps these things have to be judged against a background of the common cruelties of the time, the animal baitings, huntings, chasings, drownings, burnings, the casual and vicious ill-treatments which took place then as now – then unregulated by law, convention or social restraints.

Many anatomists appeared to relish providing an extra show at the end of a human dissection, an added frisson in the form of the vivisection of a live animal, accounts being so graphic and disturbing that, like the whispered memories from prolonged and punishing sieges, peculiarly sanguinary battles or wanton massacres, they live on to be revisited, unforgotten over time. Public vivisections had long been carried out by men who – like Galen and Vesalius – took enthusiastic part in theatrical displays of surgical prowess. Vesalius's student Realdo Colombo famously vivisected a pregnant dog, watched apparently by approving members of the clergy. The theatricality and flamboyance of the 'performance' seemed part of the attraction, the demonstration of vivisector as educator, executioner, deity, the science of the human demonstrating its alienation from the animal world in this most extreme show of human power over others.

Expressing his approval of vivisection in *De Augmentis Scientiarum* in 1623, Francis Bacon did so on the grounds that the knowledge obtained from it couldn't be found by any other means. He divided scientific experiments into '*experimenta lucifera*' or '*experimenta fructifera*' – those which simply elucidate or those which offer positive results, categories still relevant in current debates about the nature and extent of the use of animals in scientific experimentation. It is impossible to assess the effects Descartes's notion of the 'bête machine' might have had on the enthusiasm with which anatomists and scientists pursued their studies in vivisection but for those inclined to continue despite misgivings, Cartesian rationalization may well have provided a degree of comfort.

William Harvey's exposition of the functioning of the heart and circulatory system has been described as one of the most important of all medical discoveries, *Exercitatio Anatomica de Motu Cordis et Sanguinis in Animalibus*, published in 1628, confirming him as the foremost figure in the foundation of experimental physiology. Basing his work on Galen's findings, he identified how the heart operates in the circulation of the blood through extensive vivisection and painstaking work ligating and compressing veins and arteries, and analysing volume blood flow. In his biography of Harvey, Thomas Wright provides a powerful account of a demonstration Harvey gave to illustrate his theory in Nuremberg in 1636 when he – a tiny figure dressed in large white robe and cap – cut open the heart of a still-living, restrained and muzzled dog, severing its pulmonary artery in order to show the force of the flow of blood, to 'confirm his supposition', which he may have done since members of the audience were showered with the dog's blood.

Descartes, himself much interested in matters cardiac and although generally approving of the conclusions of *De Motu Cordis*, questioned Harvey's assertion that the action of the heart was that of a pump, preferring his own incorrect theory that movement of the blood was determined by heating and cooling, something he'd tried to prove by vivisection. In *The Description of the Human Body*, an unfinished work written in 1647, Descartes wrote: 'cut the pointed end off the heart of a living dog and through the incision, put your finger into one of the concavities and you will clearly feel when the heart shortens ...' and in correspondence with the Dutch physician Fortunatus Plempius, Descartes wrote on 15 February 1638 of an experiment he had carried out: 'first, I opened the chest of a live rabbit and then removed the ribs to explore the heart and the trunk of the aorta'.

Harvey's opinion of Descartes was unambiguous. In *Brief Lives*, John Aubrey recounts a conversation with Harvey, a friend, who suggested that if he wanted to read the philosophers, he should stick to the work of sound old-timers such as Aristotle, Cicero and Avicenna and avoid new-fangled types such as Descartes and his contemporaries who Harvey described memorably as 'shitt breeches'.

In his castigation of Herophilus for his possible use of humans in vivisection, Tertullian is reported to have said that Herophilus 'hated mankind for the sake of knowledge' but it is impossible to know what feelings practitioners of early, or indeed any vivisection, may have had towards their ill-fated subjects in order to allow them to carry it out. Harvey, a man not known for undue sentiment, doesn't appear to have expressed any particular disquiet about it although since he reportedly carried out a post-mortem on his own father, this seems hardly surprising. For many, knowledge was the sole and surpassing goal, one which at least temporarily overcame doubt or empathy.

Some practised vivisection without qualms, some practised it despite them and some practised it before abandoning it, sometimes in self-disgust, although often only after making the discoveries they'd sought. There were some who saw it as necessary, or even divinely sanctioned, including some of the distinguished scientists and natural philosophers who met and worked in Oxford in the seventeenth century – Robert Hooke, Robert Boyle, Christopher Wren, Thomas Willis and others whose work would lead to the founding in 1660 of the Royal Society, all of whom carried out extensive studies using animals. The polymathic Christopher Wren, architect, mathematician and physicist, conducted numerous experiments on animals, often in collaboration with the physician Richard Lower, including blood transfusions and a splenectomy on a live dog – '[I] held the spleen in my hand ... yet the puppy, in less than a fortnight grew not only as well but as sportive and wanton as before.' He also injected animals with a variety of substances, including alcohol: 'I injected wine and ale into the mass of blood in a living dog . . . till I made him drunk but soon after, he pissed it out,' he wrote. Wren was well aware of the effects of his experiments, and of others' view of them. In a letter to a friend describing them, while designating them 'cruel and horrible', he continued: 'others talk of needless cruelty. If any useful knowledge is to be obtained by an experiment, none of the means necessary to arrive at this knowledge can be needless.'

Robert Hooke, in a letter to Robert Boyle, his employer, collaborator and fellow-inventor of the equipment used in the 'air pump' experiments, wrote that he was unwilling to continue to undertake the kind of work he had carried out because it constituted 'torture', a not unreasonable description for one particular experiment which involved the opening of a dog's chest, the removal of parts of its heart and the re-inflation of its lungs, a procedure which the animal survived for over an hour. Boyle himself was heavily engaged in animal experimentation, including the injection of large doses of opium into dogs and his own experiments with his air pump using a variety of creatures, including birds and mice, all of which, he suggested with a heavy measure of self-righteousness, was his duty in exploration of God's creation. Horrifying demonstrations of some of these pieces of equipment using animals became popular public entertainments in the eighteenth century and were portrayed in Joseph Wright's painting of 1768, *An Experiment on a Bird in the Air Pump*. Sanctified though he might have considered his work, Boyle was in no doubt about its effects, describing one creature he experimented upon as having been 'furiously tortured' while dismissing questioning of his actions as 'effeminate squeamishness'. John Evelyn, the diarist and social commentator, himself a founder of the Royal Society, having watched Hooke's experiments, commented on their constituting more cruelty than he wanted to witness. The Danish anatomist and geologist Nicolas Steno referred to his own work on dogs as 'torture' and following his conversion to Roman Catholicism in 1667 abandoned his scientific studies to pursue his ministry as a bishop.

Another who expressed revulsion at his own work in vivisection was Stephen Hales, a major contributor to the field of haematology although he did so only after carrying out numerous experiments on animals. Clergyman, social reformer, physiologist, botanist and prolific vivisector, Hales obtained many of his experimental subjects as they were about to be destroyed. The best known of his twenty-five documented experiments was carried out on an elderly mare and involved inserting a canula into the horse's carotid artery, connecting it by way of a section of a goose's windpipe to a long glass tube and, on removal of a carotid ligature, measuring the height of blood levels as they altered with the actions of the heart, an important demonstration of vital aspects of haemodynamics. After the publication of his book *Haemastaticks* in 1733, Hales decided not to carry out any further animal experiments, finding them 'disagreeable', although for whom it's uncertain, and turned his attentions exclusively to botany.

In spite of the popularity of public demonstrations, there was a degree of unease about the practices of vivisection and frequent disquiet when people discovered that it was taking place. The feeling grew as attitudes towards other species began to be subject to more questioning. Hales's work, notorious around the district of Teddington where he lived, was criticized by many, including his friend and neighbour the poet Alexander Pope. A stout defender of animals, Pope wrote of Hales as 'a very good man, only – I'm sorry – he has hands imbrued with blood'. In his criticism of Hales, Pope pondered the question of whether or not we have the right to kill creatures, 'for our curiosity or even for some use for us'. In his *An Essay on Man*, begun in 1729, a poetic examination of the place of humans in a spiritual universe and endorsement of the Neoplatonic idea of the Great Chain of Being, Pope expressed his concerns:

The lamb thy riot dooms to bleed today,
Had he thy reason, would he skip and play?
Pleas'd to the last, he crops the flow'ry food,
And licks the hand just raised to shed his blood ...

The lines are reminiscent of the words of Thomas Tryon, writer, campaigner for the abolition of slavery, advocate for vegetarianism and all-round champion of animals who, in his book *The Way to Health, Long Life, and Happiness*, first published in 1683, wrote of sheep and lambs as: 'dignified with a meek and humble Nature, mild and friendly ... hardly any Creature in the World to be compared to them.' Tryon believed that sheep had 'the divine light' and were possessed of complex language with which they were able to communicate anger, hatred, love, joy and sorrow. Many writers, philosophers and thinkers of the eighteenth and nineteenth centuries, including Samuel Johnson, were in total opposition to vivisection, believing that any justification for it was invalidated by the infliction of suffering.

In his 1693 treatise *Some Thoughts Concerning Education*, John Locke proposed the view that treating animals well would encourage the appropriate treatment of humans, an idea which reflected the writings of Porphyry and others. His thoughts were arranged under such alarming headings as 'Cold Water', 'Against Tight Lacing', 'Meals', 'Sleep', 'Bowels', 'Self-denial', 'The Rod' and 'Stop Whining'. In 'Prevent Cruelty and Mischief' he says: 'For the Custom of tormenting and killing of Beasts, will, by Degrees, harden their Minds even towards Men ... and be taught not to spoil or destroy any thing, unless it be for the Preservation or Advantage of some other that is nobler.'

Locke's contemporary, the Dutch philosopher Baruch Spinoza, one of the most profoundly provocative and original thinkers of his time, rejected traditional ideas of God and nature, preferring to see God as nature, part of the substance of the universe, free from the human-imposed values which prevailed in ideas of hope of rewards, or fears of damnation, beliefs often described now as 'pantheistic'. (For his efforts, he still holds the distinction of being one of the few people ever to have been expelled from Judaism, or perhaps only the most famous, having been cast out thunderously by the Portuguese synagogue in Amsterdam for his heretical beliefs, a circumstance about which he was impressively unbothered.) He rejected the view of animals as insensate: 'for after we have learned the origin of the mind, we cannot doubt that animals feel'. His belief that 'the law against killing animals is based more on empty superstition and unmanly compassion than sound reason' is thought to reflect his profound concern that too much attention to animals would be harmful to the encouragement of good relations between humans, often explained by his deep anxiety about the human conflicts of his own times. Living quietly and ascetically as a lens grinder, Spinoza is reported, rather oddly, to have laughed aloud while watching fights between spiders caught deliberately for the purpose.

When William Hogarth published a set of engravings, *The Four Stages of Cruelty*, in 1751, they excited immediate public attention. Their message was similar to Locke's – that animal abusers invariably progress to worse. Depicting the life of the criminal Tom Nero, the engravings are horrifying, showing random and savage acts of cruelty to humans and animals carried out in the streets of London. Designed as a deterrent, they trace the degradation of the characters of abusers through the stages by which Nero progresses from violent child and animal torturer to murderer and eventually, by way of judicial process, to convicted, hanged and dissected criminal. Hogarth's message was endorsed by Immanuel Kant in *Lectures on Ethics* in which he referred to *The Four Stages of Cruelty* as 'a good lesson for children', adding, 'the more we devote ourselves to observing animals and their behaviour, the more we love them . . .' He mentioned with approbation that Leibniz, after observing a grub, replaced it on its leaf rather than do it any harm. Kant's view was that duties to animals were 'indirect', secondary to duties to humans and that the benefit to humans of vivisection, which he acknowledged to be cruel, made it acceptable as long as it was not carried out unnecessarily.

Despite the fact that it is frequently cited, the utilitarian philosopher Jeremy Bentham's famous statement on animals and their qualification for ethical treatment '. . . the question is not, Can they reason? . . . but Can they suffer' did not include opposition to vivisection. In 1825, in a letter to a newspaper, he wrote: 'I have never seen, nor ever can see, any objection to the putting of dogs and other inferior animals to pain, in the way of medical experiment, when that experiment has a determinate object, beneficial to mankind, accompanied with a fair prospect of the accomplishment of it.'

In their consideration of the extension of rights and liberties, both Voltaire and Jean-Jacques Rousseau extended their ideas to the relationship between humans and other species. Voltaire, a resolute critic of Descartes, wrote: 'What a pitiful, what a sorry thing to have said that animals are machines bereft of understanding and feeling . . .' And arguing the case for animal souls, 'The philosopher who said "*Deus est anima brutorum*" ['God is the soul of animals'] was right: but he should go further.' In his treatise *Emile*, Rousseau likened the underestimation of animal feelings and indifference to their sufferings to the ways in which the rich treat the poor: 'By extension we become hardened in the same way toward the lot of some men, and the rich console themselves for the harm they do to the poor by supposing that they are stupid enough not to feel it.' Conscious of the possibility that other species were capable of sensitivity, he wrote in *The Social Contract*: 'One animal never passes by the dead body of another of its species . . . there are even some who give their fellows a sort of burial', and in his final work *Reveries of the Solitary Walker*, he mused on the difficulties of studying an animal world whose avian, piscine and quadruped inhabitants never stay still long enough to be examined, worrying that if he had to run after any of them, he might become out of breath:

So I should have to study them dead, to tear them apart, remove their bones, dig deep into their palpitating entrails! What a terrible sight an anatomy theatre is! Stinking corpses, livid running flesh, blood, repellent intestines, horrible skeletons, pestilential vapours! Believe me, that is not the place where Jean-Jacques will go looking for amusement.

As the empirical, scientific basis for medicine was being expanded during the nineteenth century, the use of vivisection increased, as did public expression of opposition to it. Whatever the impact or influence of new ideas about the inclusion of other species in political and philosophical debate, concern over the treatment of animals began to be demonstrable in the introduction of legislation and the establishment of voluntary bodies to observe, curb and punish cruelty. Richard Martin's Cruel and Improper Treatment of Cattle Act was introduced in 1822 and in 1824 the Society for the Prevention of Cruelty to Animals established, albeit with a limited remit and the bar for what might be considered 'wanton cruelty' set very low. At the time, vivisection was carried out more extensively in continental Europe than in Britain and as a result it became a focus in Britain not only for organizing in defence of animals but for vigorous opposition to what was regarded as a predominantly foreign practice. It provided a convenient rallying point for a multi-faceted and profound xenophobia, which was given momentum by the work of two French physiologists, François Magendie and later his pupil Claude Bernard. Both men made undeniably important contributions to studies in medicine – Bernard in liver function, diabetes, homeostasis and the vasomotor system, among much else, and Magendie, in his observation of anaphylaxis, in neurophysiology and pharmacology. Both did so using equally undeniably awful experiments on animals.

The Scottish anatomist Charles Bell (after whom Bell's palsy among other conditions, was named) became involved in a prolonged and labyrinthine dispute with Magendie over a piece of seminal research in neuroanatomy over who first discovered the differing functions of the ventral and dorsal roots of the spinal nerves. (The dispute was eventually resolved by the expedient of naming the finding 'the Bell-Magendie Law'.) Working before the introduction of ether, Magendie carried out his work on puppies, Bell on rabbits, but the latter's work stalled as he began to question his own actions. In a letter to his brother in 1822 he wrote: 'You may think me silly but I cannot perfectly convince myself that I am authorized in nature, or in religion, to do these cruelties.'

Magendie though, suffered no such doubts, continuing to be adamant in his defence of the methods that had given him a reputation for exceptional cruelty, even among his colleagues. In 1824, a demonstration in London, during which he's said to have dissected the facial nerves of a dog nailed by its paws to a board who he then left, half-dissected overnight, was widely and sensationally reported in the British press. His refusal to use anaesthetics even after they became available in the 1840s, which he justified on the grounds of it affecting the results of his studies, was viewed as an aspect of an inherently sadistic nature. His pupil, the erstwhile playwright Claude Bernard, similarly protective of his methods, became even more notorious for experiments that included paralysing animals with curare before vivisecting them and studies of homeostasis that involved placing animals in a piece of Heath Robinson-esque equipment in which they were slowly roasted alive, perhaps lending credence to stories that circulated about the screaming of animals being heard in the district of Paris where his lab was situated. Repulsed by the nature of his experiments, his wife, provider of the dowry that had funded his work and early member of the Société Protectrice des Animaux, left with their two daughters to set up an animal sanctuary.

The aims of nineteenth-century anti-vivisectionists appear, on the surface at least, similar in their desire to protect animals from suffering, but their motivation varied according to their inspirations, beliefs or politics, whether in humanism, in the feminism which found identification with the ill-use and oppression of other species or, under the influence of Darwin's ideas, in the revelatory demonstrations that linked species by a common past. A fear of advancements in science encouraged a view of the questionable morality and intentions of its practitioners. These views though, didn't necessarily suggest universal identification with or sympathy for 'the other', human or not. Even with the rise of opposition to vivisection, basic notions of human superiority and sovereignty over other species (and other humans) were still too rarely subject to serious philosophical or theological challenge.

The first anti-vivisection organization, 'The Victoria Street Society for the Protection of Animals Liable to Vivisection', was set up in 1875 by Frances Power Cobbe, Irish suffragist, writer and social reformer. The previous year, she had produced a petition calling for the investigation of the practices of labs and teaching hospitals. Although broadly sympathetic to its aims, Charles Darwin refused to sign it, expressing concern about Cobbe's intemperate criticism of the eminent German physiologist Rudolf Virchow and the likelihood of the petition's success in impeding the progress of work in physiology. Virchow, a pioneer of modern pathology, the first to describe and name leukaemia and 'zoonotic' diseases, was also a liberal social reformer and vigorous advocate for improved conditions for the poor.

Darwin's stand exemplifies the dilemma of those concerned for animals, in the contradictions inherent in the life of the scientist, or indeed anyone sensitive to the interests of other species. Darwin defended the use of vivisection, which he had practised himself, as crucial for scientific development although he described his killing of one of his doves 'an angelic little Fan-tail ... at 10 days old' as 'the black deed'. Sternly critical of unnecessary suffering caused by needless repetition and the under-use of anaesthetics, he wrote to a friend in 1871: 'You ask about my opinion on vivisection. I quite agree that it is justifiable for real investigations on physiology; but not for mere damnable and detestable curiosity. It is a subject which makes me sick with horror, so I will not say another word about it, else I shall not sleep to-night.' Using his eminence and breadth of contacts, he worked on the clauses of an alternative to Cobbe's proposed bill, neither of which was passed. Appearing before a Royal Commission in November 1875, he summarized by saying that causing needless suffering to an animal was worthy of 'detestation and abhorrence'.

In 1891, Cobbe wrote to the *Jewish Chronicle* to complain to its readers: 'Throughout Germany and Austria, the great majority of Vivisectors are Jews.' Cobbe had already carried out a campaign against the highly regarded German Jewish physiologist Moritz Schiff who had studied in Paris under Magendie (while working simultaneously as keeper of ornithology at the Zoological Museum where he classified the birds of South America) and served as a surgeon in the Baden Revolution of 1848. Schiff eventually secured a chair in Florence where he worked for many years until Cobbe's campaign, during which she recruited the British expatriate community to her side. Schiff undoubtedly used animals extensively in his work but explained his position on vivisection in *La Nazione* in 1864, writing that he worked only on anaesthetized animals. His work included pioneering research into the function and treatment of the thyroid, vasomotor nerves and the treatment of cardiac arrest. Cobbe was unrelenting in her campaign and Schiff was obliged to flee to Geneva.

Cobbe's claim was and is impossible to substantiate (and if it had been, it's difficult to know what the readership of the *Jewish Chronicle* might have been able to do about it). If Jews were practising as doctors and physiologists in greater numbers in continental Europe than might be expected, it may have been because they were restricted from pursuing civil service or academic careers, limited in their opportunities by the *numerus clausus* applied in many universities, but her association of Jews with cruelty was neither incidental nor without lasting consequences.

In a paper on the rhetoric of animal rights, the historian Anita Guerrini examines nineteenth-century ideas such as Cobbe's, which connected Jews with vivisection, creating a stereotype of the Jewish scientist with origins in the same myths of blood and murder that have persisted throughout European and Christian history. She traces the idea to the influence of Schopenhauer who talked of 'Judaized despisers of animals' and Richard Wagner, who, in combining anti-vivisectionist sentiments with the vituperative anti-Semitism for which he's known, encouraged the idea of a link between Jews and cruelty in their 'polluting' of German culture. Wagner believed that within German culture there was a mystic, spiritual bond with animals – one of the ideas that was to provide foundation for the grotesque anomaly by which the Third Reich enforced strict animal protection laws while murdering millions of humans.

Cobbe didn't limit her prejudice to Jews – she was excoriatingly offensive about fellow Irishmen, Scots, Catholics and foreigners of all sorts, particularly the French and Italians. Her class prejudices were profound. She sought political 'equivalence' but not equality for women. While decrying vivisection, she defended fox hunting, hare coursing and duck shooting. She wasn't alone in her inconsistency. Montaigne himself had been fond of hunting while Richard Martin MP, also known as 'Humanity Dick', introducer of the 'Cruel and Improper Treatment of Cattle Act' and animal rights campaigner, was an enthusiastic fox hunter. Queen Victoria, herself keeper of a deer park and hunting hounds and loyal supporter of her husband's extensive hunting activities, was critical of vivisection while many well-known literary and political figures expressed views on what they perceived as its iniquity. George Bernard Shaw reported having been surrounded at an anti-vivisection rally by women in fur coats and feathered hats. One of the more colourful, if lesser-known figures of the Scottish Enlightenment, John Oswald, is best known for his pamphlet 'The Cry of Nature; or, An Appeal to Mercy and to Justice, on Behalf of the Persecuted Animals', a wild, rambling peroration against cruelty to animals and the benefits to soul and character of vegetarianism, which included flights of cringe-inducing poetic elevation such as: 'Why, she cries, oh! Why should thou dip thy hand in the blood of thy fellow-creature without cause?' Having left Edinburgh to become a soldier, Oswald eventually fetched up in Paris where, as a colleague of Danton's and member of the Club des Jacobins, his suggestion of carrying out the mass execution of opponents is said to have been greeted with the response from his friend Tom Paine, 'Oswald, you have lived for so long without tasting flesh that you have a ferocious appetite for blood.'

In 1892, Henry Salt, social reformer, socialist and teacher wrote a book which would presage later thinkers on the subject of the treatment of both humans and animals: *Animals' Rights: Considered in Relation to Social Progress*, an innovative and profound examination of every aspect of human behaviour towards other species and, indeed, towards other humans. These ideas found echo in the 'slippery slope' argument, which suggested that animal experimentation was just the beginning of a frighteningly unknowable process that might lead to cruelty without limits, a fear expressed by Tolstoy and George Bernard Shaw. Salt was deeply critical of the religious and social ideas which established and enshrined oppression towards other classes, races and species, arguing vehemently and often amusingly, against field sports, the use of animals for food and in experimentation. His work was influential and admired but didn't achieve the change in thinking that would happen many decades later.

Nineteenth-century opposition to vivisection abated with the century. By the early years of the twentieth, better health care and improvements in public hygiene had their effects in the prevention of disease and offered tangible proof of the benefits of scientific medicine (whether based on findings arrived at by vivisection or not) and within a few years there were other matters to contemplate when it came to violence, cruelty and death. By the middle years of the century, the warnings of some anti-vivisectionists, including Tolstoy, had come true. There was a slippery slope but not the one most had anticipated. If the 'Jewish scientist' stereotype had been debunked, it was comprehensively replaced and the boundaries of cruelty moved for ever. At Nuremberg, twenty doctors were indicted and seven executed for their experimentation on human subjects and other crimes against humanity, while many others who were involved, escaped and disappeared. (Non-consensual experiments on humans have been done in Britain, Canada, the United States, Manchuria, Australia and other places, both before the war and since: on the poor, the indigenous, on women, African Americans and those vulnerable to the supremacist, colonialist mentality of the relevant researchers. They were often equal in their cruelty to the experiments of the Second World War, although none were quite as extensive or egregious.)

With the exposure of the baselines of human behaviour at the end of the war, concern for other species became subsumed for a time into the welter of post-war anxieties about how humans should conduct themselves and attention turned to the obligations and responsibilities of nations. Measures to prevent and dissuade humans from turning on one other again seemed to blend seamlessly into the confrontations of the Cold War and new research into nuclear weapons and space exploration.

Memory – personal, collective and historic – filters images through time as representatives of the human capacity for cruelty to others: images of Laika, the dog sent into space by the USSR on Sputnik 2 in 1957, or Galen's monkey, Colombo's pregnant dog or the wretched brown terrier whose sufferings, inflicted during the processes of vivisection, sparked riots in London in 1907. One lurid animal experiment carried out in the Soviet Union, an attempt to create a two-headed dog, one of many such experiments carried out at the time, was photographed for the American magazine my parents happened to be reading on a plane. I was small then but remember the shock of the image, wishing I'd never seen it, knowing that the grotesque sight would return and return as it does until today. If the fate of Laika wasn't widely broadcast, it was because these things just weren't then, not the way they are now, for in whose interest might it have been for the world to know instantly that the small, stray dog, who had been the subject of great public attention and affection, died of overheating five hours after the beginning of the mission? (A piece of music I listen to often is Max Richter's 'Laika's Journey' from *Memoryhouse*, a delicate tribute to that unfortunate creature.)

One development that may have reduced concern about the use of animals in science was the increased use of rats for experimentation. Although rats had been used since the early years of the nineteenth century, and on occasion even before, there was a vast increase in their use as experimental subjects during the twentieth century. With their unfortunate and unreasonable public image, both rats and mice were, and often still are, deemed creatures unworthy of the protections and sympathy 'deserved' by other creatures such as dogs.

The numbers of mice and rats used in experimental medical, pharmacological and industrial research since then are beyond calculation. Both were among the first species, together with *Homo sapiens*, to have their genome sequenced. In the 1950s, an experiment called the 'behavioural swim test' was devised by the psychologist Curt Richter. In these tests, rats were forced to swim in containers of water from which they couldn't escape, to measure their responses and discover how long it took for them to drown. Further development of Richter's work involved 'saving' some rats while allowing the others to swim for three days until they despaired, and drowned. Richter's tests were forerunners to those carried out in the 1960s by the American psychologist, now 'self-help' guru, Martin Seligman, which involved giving electric shocks to animals to test their responses when unable to escape. The results led to Seligman's formulation of the concept of 'learned helplessness', the tendency for the development of self-abnegation in the face of difficulty. He initially used dogs but repeated the experiments in 1975 using rats: '4 experiments using a total of 159 male albino Sprague-Dawley rats attempted to produce behaviour in the rat parallel to the behaviour characteristic of learned helplessness in the dog. It is concluded that rats, as well as dogs, fail to escape shocks as a function of prior inescapability, exhibiting learned helplessness.' A further purpose was to explore aspects of human psychological conditions such as depression and to promote the still-controversial tenets of 'positive psychology'.

Sprague-Dawley rats are among the many strains of genetically manipulated rats; there are also Wistar, Long Evans, Zuker, 'knock-out', cloned, transgenic and more, all belonging to the species *Rattus norvegicus*, selectively bred for special characteristics to make them suitable for experimentation on human conditions and diseases, an idea which reflects back to us in endless sequence, our own tendencies towards folly and self-abuse and our apparent determination to ignore the feelings and rights of other species.

Experiments using primates remain common, the rationale being that primates are sufficiently similar to humans to be of use but not similar enough to encourage attention to their protection. The monkey experiments carried out at the University of Wisconsin in the 1950s and 1960s by the psychologist Harry Harlow are among the best known. Famous as the 'cloth' and 'wire' mother experiments, they too remain unforgotten, for some because of what's seen as their lasting value in demonstrating the nature of maternal bonds and needs for physical connection; for others, for the extent of their misogyny, perversion and cruelty, and for the political and social implications of their conclusions.

Harlow's studies into maternal attachment involved removing

newborn rhesus monkeys from their mothers and presenting them with wire or cloth 'mothers', to test the effects of attachment and deprivation. That much is well known; less so is that Harlow didn't invent just the 'cloth mother' and the 'wire mother' but an 'iron mother' and a 'monster mother', which stabbed the infant monkeys, blasted them with cold air or threw them fiercely against the bars of their cages to produce fear and terror. He referred to the equipment used in the forcible mating of his luckless animals as the 'rape rack' while he and his assistant Stephen Suomi devised a deep stainless-steel isolation chamber they called 'the well of despair' into which they placed an animal, leaving him or her alone for extended periods. Their 'tunnel of terror' was a chamber into which animals could be put to bombard them with lights and noise. His creation of a 'mother machine' is a chilling vision of detachment and manipulation, of operators, exercises of power, structures of control.

By most accounts a troubled, difficult, abrasive man, Harlow claimed his experiments were connected with 'love' although the sadism with which they were carried out and the facile, smutty language he used to describe them might belie it. I've seen him both praised and criticized for his refusal to use euphemisms to blunt the horror of his work. Harlow told a reporter in 1974: 'The only thing I care about is whether the monkeys will turn out a property I can publish. I don't have any love for them. Never have. I really don't like animals. I despise cats, I hate dogs. How could you love a monkey?' At the 66th Annual Convention of the American Psychological Association in Washington on 31 August 1958, Harlow gave his address the name 'The Nature of Love'. Opening with the words 'Love is a wondrous state, deep, tender and rewarding', it was illustrated by chilling photos, film of terrified infant monkeys, of hideous 'wire mothers' and ended with the suggestion that the practical application of his work might be as a response to the child-rearing problems brought about by socioeconomic conditions which 'have led the American woman to displace, or threaten to displace the American man in science and industry'. 'We now know', he said, 'that women in the working classes are not needed in the home because of their primary mammalian capabilities; and it is possible in the foreseeable future neonatal nursing will not be regarded as a necessity, but as a luxury ... as a form of conspicuous consumption limited perhaps to the upper classes.'

While people might have been able to view the use of rats, mice or primates in experimentation as justified, concern over the use of dogs, animals long seen as companions and friends of humankind, seemed to create an imaginative bridge between humans and others. In 1945, the philosopher and sociologist Max Horkheimer, a refugee from Nazism, wrote to the democratic congressman for California to ask him to support a ban on the use of dogs in vivisection, saying that the vivisections lab is: 'the practising ground of the death camp'. After accompanying a medical superior during the selection of a dog to be used for vivisection, Loren Eiseley described the penned dogs as resembling 'men in a concentration camp, who one after the other see that something unspeakable is going to happen to them'. Of the selection of that particular dog and its killing, he wrote, 'I do not know why I remember it with such pain, but yes, yes I do.'

When eventually I read my second-hand bookshop finds, they seemed illustrative of their time. Both were written during a period of challenge to accepted ideas, not long after the influence of radical thinkers concerned with the rights, capabilities and treatment of other species – philosophers Peter Singer and Tom Regan, psychologist Richard Ryder and others – were becoming widespread, providing popular inspiration for a new and often oppositional way of thinking and acting on behalf of other species. The ideas were and remain challenging and diverse, all of them examining and questioning the rights and responsibilities, the moral equivalences, duties and status of living beings in relation to one another. Richard Ryder's influential 1970 pamphlet 'Speciesism' was written in the wake of the world upheavals of 1968 when received ideas about race, politics, gender and society were being exposed, opposed and rejected. Ryder felt that in limiting itself to humans, the radicalism hadn't extended sufficiently far, writing:

> Since Darwin, scientists have agreed that there is no 'magical' essential difference between humans and other animals, biologically speaking. Why then, do we make an almost total distinction morally? If all organisms are on one physical continuum, then we should also be on the same moral continuum. The word 'species', like the word 'race' is not precisely definable.

Arguing against the infliction of suffering on animals in research, he wrote of the illogicality of claiming that animals suffer in different ways from humans since animals' possession of nervous systems analogous to our own is what allows them to be used as experimental subjects.

The philosopher Peter Singer expanded on the idea of speciesism, arguing in his seminal book *Animal Liberation* for the concept of the 'principle of the equal consideration of interests' by which the interests of all, regardless of species, race or gender should be accorded the same moral equivalence. Tom Regan, initially inspired by anti-Vietnam War protests and the progressive thinking of the time, argued in *The Case for Animal Rights* in 1983 for a fundamental change in human relations to animals, including the abolition of animal experimentation, commercial animal agriculture and sport hunting and trapping, a view which set the case for 'animal rights' in opposition to that of 'animal welfare', one of the philosophical differences between his and Singer's thinking.

The American professor of law and prominent exponent of the idea of animal rights Gary Francione, also underlines the conflict between ideas of rights and welfare in all his work, particularly notable in his

1996 book *Rain Without Thunder*. The former, he believes, affords animals the right of freedom from exploitation of any sort while the latter seeks merely to ameliorate the many situations in which they are exploited and harmed. Despite their philosophical differences, the influence of these writers and philosophers has been profound in weakening that continuing mode of thinking and in questioning the old, familiar line that divides *us* from *them*.

My two second-hand books *Animals and Man* and *Man and Mouse* seem to exemplify the polarities of debate. Vastly different in their ideas and approaches, they reflect the perennial tensions in scientific argument about animal experimentation. In *Animals and Man*, based on her 1985 Romanes Lecture (and dedicated to 'The Memory of Alix's Dog') Miriam Rothschild discussed the topics of farming, laboratories, the countryside and home, displaying aspects of her own innovative thinking on the concerns and dilemmas in the conjunctions of science and ethics, morality and practice. Sensitive to the political and philosophical changes of the moment, and praising Peter Singer and Tom Regan's books, she wrote: 'We are about to experience a fundamental change in animal-man relationships.' Although bitterly critical of animal cruelty of all kinds, she defended the use of animal experimentation: 'I have regretfully accepted the necessity for using living animals in experiments, because I believe that certain advances in medical and veterinary science and physiology, from transplantation to vaccines, depend to a greater or lesser extent on this type of investigation.' She went on to examine and condemn most aspects of the ways animals were treated in labs, suggesting ways of ameliorating the suffering involved. In this, she reflected the '3Rs' – 'replacement, reduction and refinement' – approach to experimentation suggested by William Russell and Rex Burch in 1959, which is now enshrined in the law which regulates the way animal research is carried out in Britain – The Animals (Scientific Procedures) Act of 1986. 'Replacement' is the avoidance or replacement of the use of animals by means of using such things as cell cultures, human volunteers and computer models. 'Reduction' is the minimization of animals used in any test, although not the number of tests carried out, and the sharing of data to reduce animal numbers; and 'refinement' is the minimization of pain and suffering and improvements in welfare.

They are modest enough measures and far from abolitionist in their aims but ones of which in *Man and Mouse* William Paton was scathingly critical, seeing them as a sneaky anti-vivisectionist ruse designed to 'strangle a scientist's creative impulse at birth' by means of bureaucratic intervention. In one of the vivid differences in thinking which flare through both books, Rothschild, a contemporary of Paton's, acknowledged his remarkable contribution to medicine and skill as both writer and researcher, but criticized his book for making use of 'whitewash, albeit applied in a subdued manner' for his underplaying of the use of animals 'for the sake of knowledge rather than benefit', the essential divide between Bacon's *experimenta lucifera* or *experimenta fructifera*. Rothschild wrote: 'An increase in human knowledge is not in itself a justification for experimenting on live animals.'

Paton's book, an unyielding defence of animal experimentation, outlined – compellingly – the benefits of modern medicine, the effects of advances in knowledge, the expectations of longer, healthier lives and the resulting improvements in the lives of the other species who surround us. 'Parents who had lost most of their family in infancy, and learned to speak coolly of it, could readily be equally cool about the sufferings of animals,' he wrote. His approach was brisk and scientific, but it was when he wrote of the place of other species in relation to man that Paton further diverged from Rothschild. In her overview of religious thinking, she suggested that 'the Western world seems to have accepted God's lack of compassion for the animal kingdom, and for the next 2,000 years has lived up to this version of the Image most successfully', a view with which Paton clearly disagreed. 'The question of animal experiment depends in the end on the view man takes of himself and of the animal and inanimate world around him . . .' Paton wrote. His own view was an unquestioning acceptance of the hierarchies of the Scala Naturae, and man-dominated religious ideologies of Augustine and Aquinas. Listing capacities he considered specifically human, Paton included the concept of fairness, projective sympathy and compassion for other beings, social feeling and 'reverence'. All – apart from 'reverence', which is a difficult one to demonstrate in anyone at all – are widely and observably present in countless other species.

Both books were written in the immediate aftermath of the upsurge in animal rights activism, and reflect the anxieties generated by the frequently violent 'direct action' campaigns which targeted institutions, facilities and sometimes individuals connected with the exploitation of animals for research and other purposes. If Paton was disinclined to examine the possible reasons for these demonstrations of anger, Rothschild wasn't: 'Small wonder that the deeply concerned, highly indignant public resort to violent methods – however deplorable – of exasperated protest. We are to blame. Talk. Talk. Talk. Legislation is postponed yet again. Mañana.' While the scientific community was, she said, 'appalled' at the violence of militant activists, she believed it 'sufficiently sincere to recognise the fact that without the raucous clamour of the public *nothing at all* would have been done to improve the lot of laboratory animals.' And: 'Would vivisection experiments in the UK have fallen by 14% in 1983 and scalding and burning experiments by 60% without public pressure and the media?' she asked, 'I doubt that.'

When Charles Bell wrote to his brother in 1822 to tell him of his anxieties over whether or not he was 'authorised in nature or religion' to experiment on animals, he questioned his own motivation for doing it: ' – for what? For anything else than a little egotism or self-aggrandisement; and yet what are my experiments in comparison with those which are daily done? and done daily for nothing.' In this, he was expressing an uncertainty which lingers in the fundamental questions about animal experimentation, about who does it and why and how, about what allows them to do it and where the balance might lie between human benefit and animal suffering. The suspicion of 'egotism or self-aggrandisement' is there too, lurking behind the Byzantine procedures, the uncertainties of benefit, the personal career, institutional or financial gains, which may or may not have their part in the processes of research. They're there in the appearance of scientific self-certainty, which, in defending animal experimentation, reduces questioning or opposition to mere ignorance, the idea that 'the public' doesn't understand and the suggestion that if the case for animal experimentation were properly explained, it would no longer be opposed. People wonder, as Bell did in his talking of 'Experiments . . . done daily for nothing', if experiments are being carried out, if not for 'nothing', then for purposes less heroic or promising than the claims made for them, or where the moral line can be drawn in keeping animals in laboratories even without the questions of their further suffering. Despite laws that are meant to ensure high standards of animal welfare, terrible cruelty still takes place. In 2019, evidence from an animal testing laboratory in Germany displayed shocking levels of cruelty to the primates, dogs and cats being kept for toxicity testing. In response, Dr Jane Goodall described the scenes she had witnessed at this laboratory as 'a living hell' and called for the end of biomedical testing on 'sentient and complex animals', including dogs, cats, primates and rats.

Some tests seem worryingly enduring – versions of the 'behavioural despair test' are still carried out widely. Following the published results of one set done recently, I found the usual statement that the research was approved by the relevant national committee and that 'all efforts were made to minimize the number of animals used and their suffering'. The serious ethical implications of industrial experiments were revealed by the disclosure that ten macaques were used to measure the effects of diesel fumes in research undertaken on behalf of Volkswagen in the United States in 2015. The information caused widespread international outrage, not only because of the use of macaques but because the company's involvement in a deeply duplicitous scheme to falsify car emissions seemed a further violation of the integrity of these hapless creatures.

Things change slowly. They do perhaps in response to continued and vociferous opposition to animal testing or perhaps because of the results of studies in animal cognition, which make clear that many of our previous beliefs about the limited nature of animal experience and capacity for suffering were comprehensively wrong. It may be the growing awareness of the rights of animals that encourages questioning, or anxiety over species loss and anthropogenic harms to the natural world, or all of them, which have stimulated discussion about the efficacy of the use of animals, the reliability of results based on animal studies, possibilities for replacing them and a greater concentration on finding other ways of carrying out research. It may be too because acquiring animals, dealing with the bureaucracy involved, housing and looking after them, is now prohibitively expensive – economics may turn out to be the major factor in ending animal experimentation.

One major study in 2018 assessed the costs of both animal testing and the alternatives, and concluded that many tests using animals are too slow, too expensive and too often provide misleading results. Whatever may be bringing the changes about, there are now countless organizations which exist to find ways to use imaging, cell and tissue cultures, computer modelling and simulation to replace animals in tests. In a recent study in an American journal of toxicology, the results of the use of chemical databases by 'Read Across Structure Activity Relationships' demonstrated that chemical toxicity can be predicted more accurately by this method than by the use of animal tests, millions of which are still done every year.

Recently, I saw a YouTube clip of rats laughing. I haven't had a pet rat for years and seeing it made me think of and miss the ones I had. A lot of research has been done on rats and laughter – why they laugh, what it suggests about them and us, whether it's responsive or reflexive, whether or not it's behaviour sited in an ancient, ancestral structure of the brain, something that makes them more, or less, like us. For the humble present or past pet-rat owner like me, it's simple. Tickle a rat and she or he will laugh, some more than others, just as some may be happy and some sad, some clever or less so, lively, peaceable, enthusiastic, industrious or reflective. One scientist writes that knowing that rats can laugh makes him re-think the entire question of animal suffering. Demonstrably empathetic, socially minded, cognitively able, cooperative and prone to laughter, rats often seem like a *Rattus* monument placed at a crossroads, representatives of more than the single species they are, the perhaps, and one hopes, end-point of that same line that has scrolled through time, dividing the place where limits lie for what we might or might not do to others, the place where conscience may meet Time.

As I'm going through the websites of some of the thousands of scientific organizations that promote animal research or work to find replacements for it, I see a ubiquitous image – a mouse or white rat. There's one in particular, a photo of the same appealing white rat that appears again and again, stock-photo-poster rat, it seems, for the entire field of animal testing. He or she reminds me of the white lab rat I knew once, given to my daughter Han by a friend who had rescued him at the end of an experiment. She didn't know why he hadn't been killed as most are when they're no longer of use, but she was happy to have him, an amenable, timid fellow with a torn ear where his tag had been removed and a past she couldn't interpret. He learned quickly to respond to his name but in other ways he was unlike the other rats we'd had, unable to clean himself properly, a dissonant manifestation in a species most of whose members are mutual or self-groomers of a dedication that borders on the fanatic. Han called him Elmo and cleaned him and enjoyed his company for the short time before – in the middle of his rat-thousand-day span – he died. We always wondered what had been done to him, what had altered or removed some of his essential rat-ness, and why.

In a television interview in 1987, Tom Regan said: 'We share the earth with other biographical beings ... they're not just biological entities ... they have a life that they experience, they bring this mystery of consciousness to their lives. What we do to them matters to them.' It's his word 'biographical' which gives entirety to the lives of other species, lives with backstories, territories, histories, cultures, connections, family and social relationships, rights and consciousness, the denial of which has allowed humans to do to them everything we have for far too long.

The idea has been expressed in other ways. In her delightful book *How to Live: A Life of Montaigne*, Sarah Bakewell refers to Leonard Woolf's view of Montaigne and his well-known essay 'On Cruelty'. Woolf regarded it as an important expression of the acknowledgement of individuality in humans and other species, making Montaigne, 'the first completely modern man'. After I read this, I sought out the volume of Woolf's autobiography from which the extract comes, *The Journey Not the Arrival Matters*. It begins during the Second World War; Woolf describes his own experience of the London bombings and subsequent, tragic suicide of his wife, Virginia. In the first chapter, he writes of others, human or not: 'Each and all have a precisely similar "I" with the same feelings of personal pleasure and pain, the same fearful consciousness of death, the destroyer of this unique "I".' He recalled his childhood experience of being instructed to drown three unwanted puppies in a bucket of water, his horror at witnessing their struggle to live and realization that these creatures' experience was as his would have been, that they were individuals, 'an "I"' – and that it was 'a horrible, an uncivilized thing to drown that "I"' in a bucket of water.

In 2012, a conference of neuroscientists drew up a document known as the 'Cambridge Declaration of Consciousness in Non-Human Animals' in which they asserted, among much else connected with the possibility of consciousness existing in non-humans, that: 'humans are not unique in possessing the neurological substrates that generate consciousness ... Non-human animals, including all mammals and birds, and many other creatures including octopuses, also possess these neurological substrates.'

All the time I'm walking home from my sojourn with the ghosts of convivial philosophers, I'm weighing the questions and the doubts. What do you do with troubling knowledge? What do you do with history itself? This icy evening as I pass, the granite baronial facade of Aberdeen Grammar School glows in the thin white light which illuminates the statue of one distinguished former pupil, Lord Byron, who stands in front of the school, majestic on his plinth. Another alumnus who was at the school in the 1850s, some sixty years later than Byron, was the neurologist Sir David Ferrier, discoverer of the function of the cerebral cortex, a finding described as 'one of the great landmarks of 19th century neurology'. Much of Ferrier's work on cortical localization was done at the West Riding Lunatic Asylum in Wakefield on a variety of animals. The experiments caused undoubted and terrible suffering and became known to anti-vivisectionists, including Frances Cobbe whose organization brought an ill-prepared and unsuccessful case against Ferrier for operating without a licence. The territory of Ferrier's work, the mysterious relationship between brain function and mind, gave inspiration to writers of fiction, including Wilkie Collins and H. G. Wells, the former in the utterly dreadful book *Heart and Science*, the latter in the rather better *The Island of Dr Moreau*, both of which characterized Ferrier as a looming, sinister figure of malign intent. Ferrier's work in cortical mapping allowed the development of the first surgery for epilepsy and in 1884 the first successful operation for the removal of a brain tumour.

We live by others' grace. Miriam Rothschild suggested that those of us who use eye-shadow or aftershave lotion should reflect on the Draize eye test, the standard method since 1944 when it was devised, for evaluating the safety of substances used near the eyes. To carry it out, albino rabbits are restrained while whatever is being tested is introduced into their eyes. It is painful, always prolonged and has long been regarded as a practice of consummate cruelty. William Paton too, although explaining and defending its use, was aware of its horrors, anticipating the finding of other, less invasive ways of testing, referring to what is now a widely used *in vitro* technique.

It isn't only scent and cosmetics, it's everything else, every test or procedure, everything that's done, everything that holds our lives, the lives of others, human and not, our families, friends, beloveds whose fates may lie, as do our own, somewhere along this line of history. I've always been wary of the easy apology, the post hoc regret, the empty words to find forgiveness far too late but in this case, as in most of them, there's very little else.

6

The Museum

It's like a magic castle in front of me on this cold day in late spring. I'm wandering in childhood: everything of the past, probably everything I first learned about anything, or its distant, now forgotten roots, seems concentrated here, in this one place. The magnificent red sandstone facade has the mystic quality of endurance set in memory. I've been coming here since before I can remember, brought early as part of a Glasgow childhood. Now, it's one of these indelible markers of who I am, of person and place, of *habitat* and *niche.* In a way, I'm amazed it still exists, as solid as evidently it is because for so long, at least in the time I haven't lived in this city, it's been a place of – if not dreams then origins and ideas – and appears, on this morning of falling blossom and chilly rain, pretty much the same. The places you leave attain their own reality in time, keeping years, keeping days and people, scattering into dark, industrial fogs and days of sunlight – all the broken images of childhood. It may be in places like this, museums and galleries, where once we learned the things we know. An ecclesiastical air radiated from the place when I was small, from the architecture, the huge pipe organ at one end, music from the daily recitals rising to the pitched glass roof. It might have been a place where you were expected to believe but now, seeing it all again, I know I never did.

As I walk in, I can't remember exactly how everything used to be. Impressions and sensations have lasted more than the memory of anything particular, what I liked or didn't, what repelled me or which things I knew would illuminate my life. A lot has changed since then – there was a major refurbishment a few years ago although some of the display-creatures I remember are still here: the giraffe, the elephant, both just the same. I'm overwhelmed for a second. I feel as though I should greet them: *You! After all this time – the three of us, here together again!* They stand in the way they always did, looking out with the same fixed and limpid gaze.

By the time I first knew them, the layout and display cases were no longer the same as when the gallery opened in 1920. According to museum records, they were then divided into zoology, geology, palaeontology, mineralogy and ethnology – 'the natural history of Man'. Zoology and ethnology were always on the ground floor, 'fine art' – the art of Europe – elevated to the floor above. The lower galleries had to be repaired after being damaged during the heavy bombings of Clydeside in 1941, which provided an opportunity for the ethnology displays to be changed and 'modernized'. The galleries were redesigned and the man chosen to undertake the task was Dr Henry G. Farmer, linguist, musicologist and one-time director of Glasgow's Empire Theatre whose aim was to present the displays in a way that might provide a better understanding of the 'arts and crafts of primitive peoples'. He colour-coded them into terracotta, light-green and sky-blue to provide backgrounds for the lives and artefacts of 'lower primitive', 'middle primitive' and 'higher primitive' societies. Australian indigenous people, the Saan people of southern Africa and the people of Tierra del Fuego were assigned to the first category, Melanesians, some south and east African peoples, Inuit and indigenous inhabitants of the Great Plains to the next, with Polynesians, west and central Africans and coastal indigenous North Americans to the last, as the world's human beings were given their place on this ascending, chromatic scale. Objections were expressed by the Argentinian consulate about the placing of the people of Tierra del Fuego but the displays were not changed. The layout must have been much the same by the time I was first brought here, when I was too small even to consider what they meant, or what the elephant and giraffe, or more accurately their preserved and moulded skins, were meant to tell me or whether their immobile presence might say more about us than them.

Perhaps there really is nothing remarkable about a pair of long-dead, large, southern hemisphere creatures standing on plinths on the black and white marble floor of a northern hemisphere museum. Perhaps it's only what you should expect to see anywhere natural history's displayed – exhibits in fixed glass cases, ones unconfined by glass; creatures poised or perching, large and towering, clambering on a branch, marching nowhere, flying in a painted, artificial sky. So many people must have had their first sight of wild species and the world beyond from seeing creatures just like these. The lessons may have been, and still are of great value, the collections in museums unparalleled, instructive and important, but looking beyond the creatures and their presence I wonder what we've learned this way, from husks, from dry, disconnected displays of someone else's things.

I look for the bust of James Watt, inventor of the steam engine and son, if not of this parish, then of one not too far away. It used to stand resplendent, his marble presence seeming weightily propitious but I can't find him where he used to be, here amid the surrounding cases of engineering marvels contained in all their bright and promising precision, heralding everything we couldn't have known. 'This epoch may be defined to have started about two centuries ago, coinciding with James Watt's design of the steam engine . . .' the atmospheric chemist Paul J. Crutzen said in defining his own term 'the Anthropocene' – 'the epoch of the human'.

I turn from the cases and look out of the window onto Kelvingrove Park, to the strollers and dog walkers crossing the grass, to the dripping trees, *nature morte, nature vivante*. In the later years of school, I came here with my Higher art class, half a dozen of us marched through the West End streets from our school not far away, carrying our art materials and drawing boards. 'Af tae a board meeting, girls?' a wag called once from the bus stop as we passed. In the final months before we both left Glasgow for good, my friend Ishbel and I used to come here every Saturday to drink coffee and mooch around the galleries and draw.

In his narration for *Parting Shots from Animals* John Berger talked of the human attempt to possess animals through image and its inevitable failure, 'for in the act of possessing our images you make us what we never were'. Both elephant and giraffe are silently eloquent about what they never were. Reflecting on the spirit that drove the hands and eyes of the painters of the Chauvet Cave, Berger wrote of the 'fraternal respect' with which other species were portrayed: 'Each creature', he wrote, 'is at home in man' describing other species' first role in the human imaginations as that of 'messengers and promises'. Passing between these cases of creatures, I think of the words, *messages* and *promises* and of how images and words have signified, influenced and described our relationships with other species, with insects, birds, and every other life form. It's these frozen-in-the-moment creatures, in zoos, in depictions of all kinds, in film, story, cartoon, fable, allegory, poem, in nursery rhyme and picture book, which have formed the way we see and think, the ones which have moulded our ideas and consequently, our behaviour towards the natural world.

The first true animal fable may be the story of the doomed encounter between hawk and nightingale in Hesiod's poem 'Works and Days'. Using his close observation of the natural world, Hesiod drew parallels with his own life – his sense of injustice and view of the inevitable triumph of strength. Much of his work was later incorporated, with stories from every tradition and culture, into the wide ambit of what are known as Aesop's Fables, although the eponymous Aesop almost certainly never existed. As in every other form of anthropomorphic tale, animals are used to warn, illustrate, moralize, exemplify and appear to act out the complexities of human life. The stories may be absurd, contradictory, cruel, unlikely, lacking any measured reflection of the natural behaviour of the species concerned, whether fox, ant, wolf, hare or tortoise, but have endured because they're believed, by appearing to contain some ineffable kernel of universal truth, to be a mirror to ourselves.

Not far away, between the gallery and my school, there used to be a bookshop where my friends and I went during that time when we'd begun to outgrow the books of childhood – the Alice books, *Wind in the Willows*, *The Secret Garden*, *The Railway Children*, *Anne of Green Gables*, *Swallows and Amazons*, the books still loved and read by children today. We bought anything, everything: the newly published novels, translated classics, volumes of collected poetry I bought then would join the old books of my childhood to be lost among house-moves and country-moves and time.

I keep the books my children used to read, even now, a long time after they both left home. Sometimes I go up to the top floor to dust the shelves and riffle through pages and years, from their early picture books of animals in every situation and guise, animals dressed and undressed, speaking or not, to the later ones, which for them crossed their boundary into adult reading. The nursery rhymes are still there, hiding within their comforting scansion and familiar repetition, the dark stories of blinded mice and cats in wells, dead robins and collapsing bridges. There are the early books where animals drive cars and pilot planes, where mice play the balalaika, crocodiles are taught the etiquette of dinner and tigers invite themselves to tea, *Babar* next to *The Velveteen Rabbit*, Maurice Sendak's Wild Things still rampaging and smiling their sweet, toothy smiles. Every creature's there, every species (or the more common among them) portrayed in every way, from the absurd to the frightening, the abjectly sentimental to the realistic – rabbit, rat, dove, lion, bear, elephant, badger and bee. There are some of my granddaughter Leah's books too, ones about charming pigs and ballet-dancing mice. While I'm there, I take books from the shelves, leaf through them and think again of what they taught and how. Studies of the effects of anthropomorphic portrayals in children's literature suggest that the way animals are portrayed, with other factors such as age, surroundings and personal exposure to other species, influences children's views of animals as biological entities. The conclusion of one Canadian paper is that the depiction of animals and their environments in ways which are inaccurate, distorted or fantastical, encourages a 'human-centred view of the natural world'. It's not only the images of animals per se, that may be significant, whether they're pink or tartan or blue, ridiculous, kindly, nasty or heroic, but how they appear in relation to the human world, how they're used to affect, influence and illustrate, what messages we make them impart, what promises they make.

Sitting for a while in the armchair in Bec's old room, I take out books to remind myself what Beatrix Potter's beguiling drawings promise or Kenneth Grahame's wistful characters in *The Wind in the Willows*; and then to see what *The Jungle Book* or *Tarka the Otter* might say about the world. Given that the Potter party line isn't always consistent, could the message be that life will be harsh, hard work rewarded, laziness punished and that forgetting to carry your pig licence will always be a bad idea? I wonder about the sharp disparity between the charm of the drawings and the starkness of the words, the incongruities, the disconcerting mismatch. What's the 'promise' of the *The Jungle Book*, permeated with the profound racism, colonialism and anthropocentricism, which is Kipling's lasting legacy? What's left if you take them out? I open a page to the word 'Magic' in Frances Hodgson Burnett's *The Secret Garden* – *Magic* used to describe an almost neo-pantheistic vision of the transformative power of a natural world in which animals and birds are part of a unifying spiritual force, always capitalized: 'Everything is made out of Magic, leaves and trees, flowers and birds, badgers and foxes and squirrels . . .' It's incidental that some of these magical creatures are the very species who line up dangerously in *The Wind in the Willows* to threaten the precious, already darkening world of Mole, Toad and Badger. I turn Bec's old copy in my hands and remember that this is not just a sweet tale of river-potterings, that Kenneth Grahame's anxious foreshadowing of the First World War was also an expression of his fear at the rise of socialism, crystallized through the belligerent, anthropomorphized characters of stoat, weasel, fox and ferret. (I discovered recently that, in an unlikely critique of anthropomorphic representation, Beatrix Potter wrote scathingly of Kenneth Grahame's portrayal of Toad: 'A frog may wear galoshes but I don't hold with toads wearing beards or wigs!')

I flick through the pages of *Tarka the Otter*: 'they were related to the fitches or stoats who chased rabbits and jumped upon birds on the earth; and to the vairs or weasels who sucked the blood of mice and dragged fledglings from their nests.' Is it unusual or aberrant for stoats to chase rabbits or weasels to eat mice? Henry Williamson's text is odd. It's realistically harsh yet often piercingly sentimental, the not-quite-anthropomorphic approach to the lives of other species through the relaying of their reactions and experiences by imaginative interpretation. 'The otter, patient in life after many sorrows and fears . . .', 'he licked her face, while his joy grew to a powerful feeling . . .'. Otters may well feel these things but we'll never be certain how or what they feel. 'What is it like to be a bat?' Thomas Nagel famously asked, questioning human ability to enter the consciousness of another.

Watership Down is here, and *The Chronicles of Narnia*, both profoundly anthropomorphic. Richard Adams's rabbits are used to play out the all-too-human themes of environmental degradation and cruelty to animals, while Lewis's creatures, real and not, from albatross, bat and centaur to kraken and hamadryade, populate a disparate, judgemental world of convoluted religious allegory. Adams always claimed, disingenuously, that his narrative was a faithful reflection of the lives of rabbits, but his rabbits occupy their places as obediently as any human involved in any male-camaraderie-in-the-face-of-adversity, public-school, Boy Scout or military story. The book may be an excellent critique of the damage humans inflict on the natural world but it's also notable for Adams's uneasy treatment of sexuality, his male characters fully realized, his female ones mere shadows, there only for the purposes of breeding. The dialogue and events surrounding the kidnap of two female rabbits are dispiriting: 'are they any good?' one male rabbit asks another. After the loss of a female, a male rabbit says, 'What's a doe, more or less?' Adams's explanation was that, 'fidelity, romantic love and so on . . . are unknown to rabbits' but so are casually misogynistic dialogues.

In 1959, C. S. Lewis touched on the way creatures are represented in writing or on screen in a letter he wrote to the BBC producer who, having successfully adapted the Narnia books for radio, now wanted to adapt them for television. Lewis declared himself adamantly opposed to the idea on the grounds that animals transformed from story to image, 'turn into buffoonery or nightmare'. 'A human, pantomime, Aslan wd. be to me blasphemy,' he wrote, Aslan the lion often being considered as a representation of Christ. He might have considered any portrayal of Aslan different from his own as 'blasphemy' but since their publication, his books have been widely criticized for the racist, Islamophobic presentation of the dark-skinned 'Calormenes' and the pinched misogyny that gives girls a lesser place in the narrative and condemns the blameless Susan to hell for the sin of preferring 'lipstick and nylons' to spiritual fantasy lands. His animals are used in the playing out of his grand narrative of good and evil, made to act out the moral failings and betrayals which are, after all, an area of human specialization. After Lewis's death in 1963, the *Chronicles* were adapted for television and film, the latter by Disney, who, in his letter, Lewis had described as combining 'so much vulgarity with his genius'.

As I walk through the galleries, elephant and giraffe behind me, I think of some of the words used most commonly to describe them and their living and breathing counterparts. 'Iconic' and 'charismatic' are words which, like image, make them what we want them to be, take them from what they really are. What is an 'iconic' species? Who is 'charismatic' and who is not? Do the words confer our admiration or special protection? Do they grant us further freedom to exploit them, to use them and their images for the furtherance of our economic gain? Why this one and not that? The words we use of other species describe processes of exploitation and division, them and us, as we create aggressors and victims, innocent and guilty, substitutes, scapegoats, receivers of just punishment, the rightly, or wrongly, judged and sentenced.

'Cunning', 'wicked', 'killer, 'ruthless' are commonly used about other species: wolf and fox, gull and rat. These are words which remove each animal from the context of their lives, a means of conveying more than we wish to appear to say. A few years ago, a very ordinary altercation between two doves, released by the Pope from a window in the Apostolic Palace and a passing crow, was described in news reports as: 'Pope's white doves of peace attacked by black killer crow'. A cousin talked recently of what he'd seen years ago during the making of a public information film, the kind that used to instruct us on how to cross the road or hide under the table in the event of a nuclear attack. He's still troubled by it. Designed to warn householders about the dangers of lax home security, it used a number of very young and frightened magpies to appear to enact the behaviour of human thieves. In chaotic scenes, the terrified, panicked birds, who had clearly been released into the room, appeared to pick things from open drawers – cheap jewellery, clothing – knocking over scent bottles, smashing china. 'Don't let them get away with it!' says the clipped voice-over, 'fit proper locks!' My cousin was working on an unrelated project in a studio next door and watched as sacks containing unknown numbers of dead birds were removed during the filming. 'It was meant to be a secret,' he said, 'but everybody knew.'

'You make us what we never were,' Berger wrote in *Parting Shots from Animals*.

Magpies are not 'thieves'. 'Theft' is a human construct, a human moral idea. 'He is a merciless tyrant, a meaningless murderer,' Sir Peter Jeffrey Mackie wrote of weasels in his guide for gamekeepers in 1910, a time when the wildlife of Britain had been savagely depleted by the hunting habits of the landowning class. Our words make creatures into thieves, savages, killers, terrorists: 'Magpies are killing off our songbirds,' a newspaper says erroneously. 'Killer seagulls terrorise town.'

Through a window, I watch a magpie picking at the grass of the park. I was given a fledgling magpie once. He was with me for six years, the brightest thing I've ever known. There are no squirrels to be seen this morning. I don't know if there are any here but at home, the grey squirrels have gone, extirpated by the campaign to remove 'invasive species'. For years, we've talked about other species as strangers: 'The little owl is an alien from other shores ... undoubtedly the worst owl we have ...' H. Mortimer Batten wrote in *Woodlore for Young Sportsmen* in 1922. Newspaper headlines warn: 'Alien grey squirrels spread disease ...', 'Diseased grey squirrels are foreign imports ...', 'Alien invaders as insect numbers soar ...' but time is all that counts. How long has it been here? How long have any of us? We – humans, birds, every other species, seeds, plants, spores, have moved from place to place, all of us travelling, flying, ballooning as spiders, blown by the wind, carried on the tides, on the feet and wings and feathers of birds, as ballast or cargo, transported in every way, by every means. We've taken everything else with us, knowingly and unknowingly, for every purpose, economic, practical and sentimental. Over time, seeds spread and plants and funguses proliferate. Roses grow now where they didn't once, raspberries and blackberries, rhododendrons and euphorbia.

In the first century CE, Pliny the Elder talked of the overpopulation of rabbits in the Balearic Islands. The map of the spread of rats through shipping is a map of human exploration. *Homo sapiens* is the most invasive species there has ever been. There are species that pose dangers to the environment but the language we use poses more – the language of warfare, of swarms and hordes and tides and floods, dictated by a way of thinking that sees particular species, whether wolf or eagle, red squirrel or lion, as representing 'us'. 'Native' is desirable, 'non-native' is not in the forever narrowing borders of the nationalistic mind. In his book *Invasion Biology*, Mark A. Davies examines the terms used in this area of biology, expressing regret at the adoption of military language, which, while possibly attracting 'a group of highly motivated supporters', also encourages a 'strongly confrontational approach'. Stephen Jay Gould writes critically of the idea that 'native' species are invariably best for their situations, pointing out that: 'organisms do not necessarily, or even generally inhabit the geographic area most suited to their attributes.' He dismisses as 'romantic drivel', the idea that there is any intrinsic superiority of the 'native' species over the more recently arrived. 'How easy,' he writes, 'the fallacious transition between a biological argument and a political campaign.'

In an online forum of essays on the natural world, a woman describing herself as a naturalist writes in detail of her act of killing fledgling sparrows. She does it she says to preserve the native American bluebird from these birds who 'outcompete those who are supposed to be here', birds who 'steal' others' places. This particular argument is old and lost but still she writes of 'meting out justice' and her anguish at the onerous nature of her duty to kill. It is described moment by moment, her pulling the fledglings from their nest, dropping them into a neighbour's pond, watching the 'flapping of naked half-formed wings as the little ones begin to sink'. 'They don't belong here,' she says and asks if the bluebirds of America were to go elsewhere, would they be welcome in a place where their plumage wouldn't fit in and the notes of their song would sound 'foreign'? 'I am an animal lover,' she says.

'I am become Death, the destroyer of worlds', Robert Oppenheimer said, quoting from the *Bhagavad Gita* after the Manhattan Project's Trinity test.

'Animal lover' is one of the few terms we have to describe how we may feel towards other species. They're all inadequate, unwieldy or remote, 'animal loving', 'biophilia', 'theriophily'. 'Animal lover' is a twisted accolade, a flip-sided paradox. If we say it about ourselves, we may be delusional, sentimental, self-regarding, lacking in self-awareness – by giving the appearance of loving too much we may well love too little. Said of anyone else, the term is sly and insulting. 'So-called animal lovers' is the favoured term for those who try to justify one or other aspect of cruelty against other species, as a sneer aimed at those who are deemed to make themselves weak, 'crazy' or vulnerable by compassion. The term's meaningless. Who among us, or them, might be the beneficiary of this love? The ugly? The 'cute'? The biddable? The domesticated? The ones we deem worthy? The ones we know? The ones we don't?

The term 'theriophily', also known by the snappy term 'animalitarianism' was first used by the American philosopher George Boas who employed it in an essay in 1933 called 'The Happy Beast' where he traced this 'complex of ideas which express an admiration for the ways and characters of the animals' through folklore and philosophy. The belief that animals have qualities that challenge and outstrip those of humans can be found throughout history, wonderfully demonstrated in an exchange between Odysseus and Gryllus in Plutarch's witty dialogue on notions of animal rationality and moral superiority. One of Odysseus's men, transformed by the enchantress Circe into a pig, Gryllus – 'Porker' – is invited to tell Odysseus why he'd prefer to stay as a pig rather than be turned back into a human. Having effected the introduction, Circe diplomatically withdraws to allow Gryllus the freedom to express himself as he chooses, which he does with considerable rhetorical skill and elegance, pointing out inconsistencies in Odysseus's thinking while elaborating why animals are superior: 'Beasts never beg or sue for pity or acknowledge defeat: lion is never slave to lion, or horse to horse through cowardice . . .' Arguing that animal behaviour demonstrates superior courage, good sense and freedom from vanity, Gryllus says, 'since I have entered this new body of mine, I marvel at those arguments by which the sophists brought me to consider all creatures except man irrational and senseless.' His observation: 'I do not believe that there is such a difference between beast and beast in reason, as between man and man' was subsequently quoted by Montaigne and others. Boas relates a growth in theriophiliac ideas through history to the expansion of knowledge of science and consequently of other species in the eighteenth century, quoting as an example of later theriophily Walt Whitman's 'Song of Myself':

> I think I could turn and live with the animals, they are so
>   placid and self-contain'd,
> I stand and look at them long and long.
> . . .
> They do not make me sick discussing their duty to God,
> Not one is dissatisfied, not one is demented with the
>   mania of owning things,
> Not one kneels to another, nor to his kind that lived
>   thousands of years ago . . .

Another term is 'biophilia', an idea enshrined in E. O. Wilson's 1984 book *Biophilia: The Human Bond With Other Species*. In it, he wrote of gazing over the forests of Surinam, a memory which intensified over the years, his emotions on remembering it altering to become 'rational conjectures', which he described as: 'the single word biophilia, which I will be so bold as to define as the innate tendency to focus on life and lifelike processes'. His belief was that early evolutionary experience of the savannah has imbued in humans a sympathy with and desire for certain landscapes and connections with other species. The word 'biophilia' has been widely used since, most notably by Björk for the title of an innovative album, documentaries and projects on the natural world, some in collaboration with David Attenborough. As defined by Wilson, the word occupies an ambiguous place in the terminology of man's relationship with the natural world, perhaps because it's a concept that's difficult to prove, based on imprecise terms – what is an 'innate tendency', and what are 'lifelike processes'? – and perhaps because in the light of the evidence of species extinctions and environmental damage, there's a clear possibility that a corresponding but opposite tendency may exist.

The word 'biophilia' though, was not Wilson's invention. The first use of the word was by the philosopher and psychotherapist Erich Fromm in his books *The Heart of Man* and *The Anatomy of Human Destructiveness*: 'biophilia is the passionate love of life and of all that is alive. It is the wish to further growth, whether in a person, a plant, an idea or a social group. The biophilous person prefers to construct rather than to retain. He wants to be more rather than to have more.' Fromm's thinking was integral to the group of German, mainly Jewish, Marxist philosophers known as the Frankfurt School to which he belonged, the best known of whom were Theodor Adorno, Walter Benjamin, Max Horkheimer and Herbert Marcuse. All of them except Benjamin, who committed suicide in Spain in 1940, fled to the United States to escape fascism. Profound questioners of man's social and political relationships with one another and with the natural world, through their work they all expressed personal experience of the consequences of the misuse of power. In 'The Skyscraper', Max Horkheimer wrote of the structures of capitalism as a dizzying hierarchy of vertical exploitation, from magnate and landowner downwards to employee and provider of services, skilled to unskilled worker to the unemployed

and the poor, still further down to the then still colonially ruled workers of India, China and Africa to the lowest tier of all – exploited and
suffering animals. 'The basement of that house is a slaughterhouse, its
roof a cathedral, but from the windows of the upper floors, it affords a
really beautiful view of the starry heavens,' Horkheimer wrote.

Theodor Adorno had an affection for animals from childhood
and an unexpected fondness for giving friends and family whimsical
animal names – gazelle, hippopotamus, tiger. He composed a song of
welcome for his friend Max Horkheimer called 'Mammoth' and wrote
to the director of Frankfurt Zoo in 1955 to urge him to acquire – randomly enough – a pair of wombats, a babirusa pig and a dwarf hippo.
Always conscious of the dangers and possibilities of the destruction
of the natural world, he wrote of the prospect of society becoming 'a
vast joint stock company for the exploitation of nature . . .' accurately
forecasting the future commercialization and commodification of wild
land into 'nature reserves'.

Herbert Marcuse was one of the notable figures of the new radical
politics of the late 1960s and 1970s for his opposition to the war in
Vietnam and support of the feminist and liberation movements. He
was outspoken in linking environmental damage to the ravages of
capitalism: 'The fight against pollution is easily co-opted. Today, there
is hardly an ad which doesn't exhort you to "save the environment",
to put an end to pollution and poisoning . . . In the last analysis, the
struggle for an expansion of the world of beauty, nonviolence and
serenity is a political struggle . . .' he said in a talk he gave in 1972.

I walk up the stairs. The roughness of stone banister is familiar
under my hand, the paintings too, Dalí's *Christ of Saint John of the
Cross*, Rembrandt and Samuel Peploe and E. A. Hornel. The idea that
culture, music and art are redemptive, preventive, 'civilizing' is dubious, questioned in Adorno's *Negative Dialectics*, where he reflects on
the failure of 'philosophy, art and the enlightened sciences' to prevent
the depravity of war. In a lecture he gave in Philadelphia in 1882,

Oscar Wilde suggested that to understand art is the best way to learn to love nature: 'And, the boy who sees the thing of beauty which is the bird on the wing when transferred to wood or canvas will probably not throw the customary stone' but it is not necessarily the case. Wilde may or may not have known of John James Audubon's fondness for killing. Best known for his 584 paintings of the birds of America, all of whom, and many more, he killed in the process of fulfilling his mission to paint them, Audubon delighted in the slaughter, declaring himself disappointed if he had not killed more than a hundred a day. The well-known ornithologist John Gould, producer of beautiful, illustrated collections on ornithology, including *Birds of Europe* and *Birds of Australia*, was described by his wife Elizabeth as 'having already shown himself to be a great enemy of the feathered tribe having shot a great many beautiful birds and robbed various others of the nests and eggs ...' (It was Elizabeth who illustrated most of the books although John was given and accepted the praise and acclaim. Her reward was having the Gouldian finch, *Chloebia gouldiae*, and 'Mrs Gould's sunbird', *Aethopyga gouldiae*, named after her.)

Where do the lines fall between how we see or portray other species and what we do to them? A while ago, I was visiting another city and went to a talk on ecology and art. It was mid-December, a day promising snow, the light casting an odd, metallic glow over old stone streets. The event, in a gallery attached to a university, was held in conjunction with an exhibition on the use of natural forms in art, 'at a critical time', which included depictions and explorations of landscape, multidisciplinary studies of sea and river water and examples of taxidermy. The talk was to begin in early afternoon – each artist was to explain their work, their inspiration, the ecological, artistic and philosophical bounds of their ideas. Beforehand, I wandered among the photographs, paintings and videos and looked at rows of once-creatures displayed on plinths – birds and mammals, none readily identifiable as who they'd been. The audience assembled and the artists began their talks, the first about water as representation and resource. The images were beautiful but terrible, the words those of rising, melting, drowning. I watched through the huge skylights for the snow to begin. Somewhere in an old part of the building behind us, a choir was practising for a Christmas concert, sending voices spiralling through empty stairwells. Landscape followed – hot, desert vistas on a day of cold – winds and dust and heat. These, the artist said, are the contradictions and symbols of our times and he gestured upwards as the snow began, a quick, sudden swirl out of a grey and violet sky. Outside, gulls flew easily through the thick, grey flakes, underlit in diminishing daylight as they touched down above us, big, soft feet sliding on sloping glass.

The final talk was on taxidermy and we were told about the use of the bodies of dead animals to make art, ideas of cultural representation, the value of the shocking and its role in helping us delineate and understand our relationship with nature and other creatures, obliging us to face, through the medium of 'materiality', the prospect and inevitability of death. Verisimilitude, the artist said, was not her aim. Then she showed a brief film of her working method and at the sight of a cat's skin being stripped back, a young woman in the audience stood up, her face waxy white. Someone helped her out and the mood of the audience in the gallery, which had been receptive enough, seemed to change to become restive and faintly hostile. An elderly gentleman suggested quite forcefully that what the artist was doing was a betrayal of art. The organizers stepped in hastily and called for questions and then it was over. The doors were opened and the audience rushed out to the freedom of the twinkling, Christmas-lit streets and the gorgeous inconsequential chatter and the comfortable seats of nearby coffee houses. Behind us in the gallery, the paintings and photographs hung against blank white walls. Sandstorms gusted and swelled and smothered. Water surged and lapped, surged and lapped and the disassembled parts of once beautiful, living animals and birds stood on their plinths like sorry memorials to nothing at all.

Travelling home that winter day, I thought of those images. Looking down from the plane window, I remembered Lynn T. White's words about the human view of ourselves in the 'natural process', our distance from it, our willingness to use it, and of our 'contempt', and wondered again what 'the natural process' is now and what constitutes usage, and what contempt really is. Below us, snow skimmed the surfaces of the land, thinly at first, thickening as we flew north. Watercourses glinted, flickering under blowing clouds and I tried to imagine what those creatures I'd seen had been and where the difference lies between image and the probably brutal actuality of violence-made-presentable in using the dead of other species for display, trophy, art.

Taxidermy began with the Egyptians. The first known and prolific preservers of bodies were, through their acts of preservation, expressing the never-ending human desire for the continuity of life, life after death, life after life. Animals might be sacred, scarab beetle, jackal, bull, crocodile, cat or ibis, companions in a future life. Preserving skin and flesh was and remains difficult to do but 5,000 years ago they did it with conifer resins, aromatic plant extracts, fats and oils, entombing their dead with artefact and provision, with plant-fibre sandals, jewellery and arrows, with amulets, knives and glassware, mummified mice, woven baskets and ostrich-skin bags.

Over time, the purposes of preservation changed, animal hides were tanned and cured and worn, entire animal and bird bodies preserved to be used as objects of study, decoration, trophy or one day to be added to the indulgent bizarreries of the *Wunderkammer*. After the depredations of colonialism broke open human and animal worlds to plunder and exploitation, everything became subject or resource, slave or loot, the bodies of other species, as well as those of humans, turned to symbols of individual or national power in a long, triumphalist display of conquest and possession.

In her essay 'Teddy Bear Patriarchy: Taxidermy in the Garden of Eden, New York 1908–1936' the anthropologist Donna Haraway

174

traces relationships of human and natural worlds through the dioramas in the American Museum of Natural History in New York.
Inspired by the grandiose male visions and aspirations of Theodore
Roosevelt and created by the hunter and taxidermist Carl Akeley,
the formalized displays are of animals in cases set against painted
backgrounds. For Haraway, they represent narratives of historic and
contemporary culture in their interplay of power in gender, colonialism, race, urban and 'wild', what she describes as 'civilization and the
machine'. Although writing specifically of one museum, her suggestion
that relations of power emanate from such images, applies to many
more. Roosevelt, renowned as a conservationist and lover of nature,
expressed his concern about the need to preserve wild creatures: 'Let
us hope that the camera will largely supplant the rifle,' he wrote, but
returned from his hunting trip to the Congo with Akeley in 1910 with
11,788 slaughtered animals. An event during one of Roosevelt's hunts
provided the origin of the name 'teddy bear'. During an expedition
in 1902, faced with an old black bear which had been tied to a tree
for him to shoot, Roosevelt instructed that it should be 'put out of its
misery', which it then was, reportedly with a knife. This is a rather
different account from the popular one that suggested he had spared
the life of a baby bear. Haraway describes Carl Akeley's lifelong striv-
ing to find entire families of the largest and most perfect animals to
shoot, bring them back and reincarnate through taxidermy, in sublime
imitations of themselves. Set in picture-perfect representations of the
places from which they originated, these creatures have become, as
Donna Haraway suggests, more than just themselves, in their endless
reflection of the destructive, manipulative power of man.

The thought of those dioramas make me remember J. D. Salinger's
young protagonist Holden Caulfield in *The Catcher in the Rye* remi-
niscing as he walks to the American Museum of Natural History. He
recalls his childhood visits to see the glass cases of deer drinking at
waterholes, the stuffed birds on wires still migrating south, the 'pottery

and straw baskets and all the stuff like that' believing that the best thing about them was that they were always the same and it seems that it is his fear of facing the difference in himself that prevents him from going in when eventually he reaches the doors. I read the book just before I left Glasgow and now in this gallery, like him I think of difference and change.

Downstairs here, there used to be a case containing a Lakota Ghost shirt bought by the museum together with a warrior's necklace, a boy's moccasin and a baby's cradle, from a man working in Buffalo Bill Cody's Wild West Travelling Show, which spent the winter of 1891–2 in Glasgow. It was a year after the Wounded Knee Massacre where the shirt, an important and symbolic spiritual item, was plundered from its dead owner. I remember the shirt, its name and resonance more than the object itself; one of those casual acquisitions, in the easy proximity of cultural relics, the precious, the sacred, the day-to-day, the animals on plinths.

The attitudes which allowed the looting of things filtered into their displays in the choice of placing, in the human zoos, the 'ethnic' shows, the 'colonial exhibitions', the 'ethnological-missionary exhibitions', the vaudeville displays put on by P. T. Barnum, the Buffalo Bill Wild West Shows. They were all designed to reinforce and vindicate the ethic of plunder and murder, the spectacles of humans and animals together, unified in their 'exoticism' and servitude. Carl Hagenbeck, a German animal dealer turned impresario, trained some of the astonishing numbers of animals he arranged to have hunted and captured from African countries to perform in his circuses and the 'ethnographic' shows for which he also imported people from Lapland and Sudan.

During the 1880s and 1890s, the photographer Roland Bonaparte amassed a vast collection of his own photos of people brought by force to Europe from Australia, North America and Lapland. These were mainly portraits and family groups. The images cannot obscure either the manifest callousness of the enterprise or the obvious and terrible suffering of the subjects, many of whom, captured and enslaved, died from tuberculosis, smallpox or flu, many never able to return home. Bonaparte was a student of the neurologist Paul Broca (most famous for having part of one of the frontal lobes of the brain, 'Broca's area', named after him) a man devoted to the study of skull sizes and the other physical measuring used in the sinister endeavour of the 'scientific racism' that allowed and encouraged such cruelty. Here, as in most museums, 'biological material' was collected and kept, including human remains, Australian aboriginal skulls (since given back, as was the Ghost shirt), part of the degrading backwash of empire whose boundaries, topographical, cultural and moral, were always fluid and unreliable, underpinning the ideas of Broca, Hrdlička and others.

If in time, human zoos became obsolete and discredited, animal zoos did not. Carl Hagenbeck, the man who had, with great cruelty and disregard for animal life, traded thousands of lions, rhinos, bears, tigers, giraffes and other species and made humans and animals into spectacles for the entertainment of others, entered the increasingly lucrative enterprise of animal zoos. In 1874 he created the Tierpark in Hamburg, which he later redesigned with moats and trenches instead of cages, patenting the design in 1896. In *Minima Moralia* Theodor Adorno wrote of zoos as a paradigm for the human state, 'allegories of possibility', symbols of the decline of civilization deluding us into believing that we might avoid inevitable destruction. 'Zoological gardens in their authentic form are products of 19th century colonial imperialism ... The more civilisation preserves and transplants unspoiled nature, the more implacably the latter is controlled,' he wrote. Like Berger who considered even the visibility, air and space as mere tokens of freedom, Adorno too viewed the open layout of Hagenbeck's designs as no kinder than that of cages, the invisibility of boundaries serving only to reinforce the completeness of the animals' imprisonment.

Sometimes, in darkness, I try to recreate in my mind the night sounds of a zoo. I remember London evenings in the flat where my mother lived beside Regent's Park after we left Glasgow, the cries from the zoo calling into the late still air, the voices of unknown creatures in their strange and lonely translocations. One path through the park ran beside the cage for wolves and I often walked beside them as they paced, the wire between us, a wolf accompanying me until I had to walk on without her because there was nowhere further for her to go.

After the talk on taxidermy, I thought a lot about the imprisonment of animals in actuality or image, of its renewed, and it seems lasting, popularity. We're all familiar with old taxidermy, with the elephant and giraffe, the kind found in stately homes, in Aberdeenshire castles whose turreted roofs you spy through trees at the end of long drive-ways, the kind in dusty halls, hung on baronial panelling, walls of skulls and antlers, heads of deer and stag, sad-eyed antelopes, eland and kudu, impala and gemsbok. We've all looked into cases of now decaying pets, at dust flying from the scruffy fur of unidentifiable mustelids, at elephant's amputated feet or the moth-nibbled wings of once-fine birds and we may all have felt the same inexpressible sadness at what they have become.

'Taxidermy,' Rachel Poliquin says in her book *The Breathless Zoo*, 'stops time'. It stops time but magnifies it too. It refracts everything we want to say about ourselves and others. It is often not in any present cruelty – the cruelty may have taken place long ago or not at all. Often, it's delegated cruelty, the cruelty of others. It is not even the macabre nature of it that speaks, it is what the images say and what they represent. A magazine for antique buyers discusses the market for taxidermy, writing of it as 'a respectable interest' while advertising Victorian specimens of great bustards and spoonbills. They mention that when the particular specimens of great bustards (*Otis tarda*) were killed, the birds were almost extinct in Britain, the last having been shot in 1832. They are, they say, being reintroduced in southern England. In *Silent Fields*, Roger Lovegrove wrote: 'It is a sad indictment of the past abuse of wildlife in Britain that one of the priorities for wildlife conservation has to be the recovery of species that were deliberately eliminated in previous centuries ...'

Another dealer sells one of the tableaux of Walter Potter, the English Victorian taxidermist, best known for his recreations of anthropomorphic scenes. His was not the scientific taxidermy of the great museums of the time, the desire to preserve and learn, it was the exercise of whimsy. 'Morbidly cute', someone writes of them, 'morbidly adorable'. But badly stuffed kittens, mice, rats, frogs, birds and squirrels acting out faux-human weddings, schoolrooms, tea parties, croquet games, the participants dressed and manipulated into grotesque masquerades, seem little more than warped control disguised as entertainment.

In a magazine designed for the ultra-wealthy, an advert appears for the jewellery department of a well-known department store. Designed by a popular advertising agency, it features a taxidermied northern bald ibis standing on a square white plinth, wearing a necklace made from the rare gemstone tanzanite and many diamonds. The advert is described by the advertising industry press as 'tongue in cheek' and 'playful'. At the time of its design and publication, the northern bald ibis (*Geronticus eremita*) was fast becoming extinct in the Middle East, hastened by the war in Syria. A few birds were found near Palmyra in 2002 but by 2014 only a single bird was left. Now, in a region heavily affected by the presence of Daesh, the bird is almost certainly extinct. This magnificent, although undeniably odd-looking creature, has suffered its decline, as most other species, through the activities of man. Under the photo of the bird in the necklace are the words 'Dead Rare'.

On an interior decor website, stuffed grey-crowned cranes are advertised. They are birds on the International Union for the Conservation of Nature list of 'red-endangered' species. How did they become an *objet decoratif*? Did these ones – as we're assured – die naturally? They're sold as matching pairs. Two contributors to the American taxidermy exhibition suggest that artists should 'use animal bodies that are already dead' whenever possible and must be both 'able to speak about the death of the animal' and willing to face the choices that they have made.

The fashion for taxidermy often seems like a passing phase that doesn't pass, a continuous revival, an 'ironic' gaze back to colonialism, retro, hipster, an untimely social phenomenon, subject to the zeitgeist, the moment, to *forces*, postmodernism, post-humanism, post-structuralism, lacking satisfactory explanation about why at this moment we – or some of us – feel it appropriate or rebelliously inappropriate to keep dead, stuffed creatures in attitudes natural or unnatural in our homes or display them in our restaurants at this time when increasingly we must be concerned about what we eat and what we wear and how we live and the frightening depletion of species in a warming world.

'Taxidermy animals are extraordinary animal-things,' the exhibition notes from a large American art exhibition say. They may be. They may be 'pre-expired', 'zoo waste', 'road kill', 'natural', 'botched', 'animal things' or 'animal objects'. They may be beautiful, meaningful, declamatory, vulgar or tacky or they may be shocking, ugly, distorted, hugely lucrative or righteous, angry declamations against the misuse of animals and the degradation of their and our environments. They may, as some suggest, be a way of deepening our understanding of our own prejudices and perceptions of other species, as philosophers have tried to do in reconceptualizing our relationship with other species through enquiry and ideas, but they may just be the bodies of dead animals.

The idea that taxidermy is instructive to those of us who might not be aware of the implications of mortality is dubious – lessons in mortality may be learned better from Seneca than from rotting cattle in leaky tanks. In a wry article about the number of animals killed for inclusion in Damien Hirst's many works, the website Artnet News, with admirable effort, calculated it at 913,450. Describing Hirst as 'thanatotic' – having an instinct toward death – they quote him in conversation with the writer Gordon Burn: 'Cut us in half,' Hirst said, 'and we're all the fucking same.' We may be, but for ourselves we observe long-lasting taboos, demands for respect, reverence, the correct treatment of the dead, which instil in some of humanity at least the belief that there are limits, however sparsely observed, in the things we may do to other people, living or dead.

In counterpoint to his idea of 'biophilia', 'a love of life', Erich Fromm developed the idea of 'necrophilia' as an attraction to death and the dead, not in a sexual sense but in a way that reflects a destructive urge to turn the 'organic into the non-organic'. In 'Creators and Destroyers' he examines the idea, relating the negativity of 'necrophilia' to ideas of control, possession, destruction and commodification through the medium of death.

The taxidermy classes given by the artist whose talk I attended are heavily over-subscribed. I look at her website and wonder what expression of ourselves it may be when it is characterized by a badly stuffed mouse in spectacles. Someone else makes earrings from dead mice. Online advice from an American organization about hunting rituals and how to treat your taxidermied 'kill' is clear: 'You should never desecrate head mounts by placing cigarettes in their mouths, sunglasses over the eyes or Santa Claus caps on their heads ... That dishonours the creature. It's a matter of respect,' it says on a National Rifle Association website for hunters. The difference has been made between animal as trophy and animal as artwork, but in the use of an animal body, there is nothing more powerful than the living, more tractable than the dead.

'I hate museums; there's nothing so weighs upon my spirits,' Henry David Thoreau wrote in an undated diary entry, complaining of their distance from nature, from the 'green bud of spring ... one faint trill from a migrating sparrow ...' what he called 'dead nature collected by dead men'.

I wander through the displays of stained glass, the dim galleries glowing with points of clear red and blue. Light glimmering through lozenges of turquoise and emerald turn the marble floor into the fluid green of wet grass and rock pools. Glasgow is famous for its stained glass, much of it domestic, like the long window facing the stairs in the house on the other side of the city where I was brought up, a Dutch scene executed in art nouveau style, a woman and child walking into a landscape, polders, a windmill, a canal.

Downstairs in the Egyptian galleries, I remember an odd and salutary story. Mummified creatures were not all the same, some being more significant than others. There were 'cult mummies', animals who were well treated during their lives and buried ceremonially after their deaths and mummification; the mummified pets and the 'victual mummies', those destined as food for the afterlife and the lower rank of mummified creatures, the votive mummies, used for religious purposes similar to the use of candles in churches. Votive mummies were required in such large numbers that it has been suggested that certain birds and animals were 'farmed' – bred specially for the purpose of being killed to be made into votives. In 1889, a huge burial haul was found in Egypt, at Beni Hassan, near the shrine of Speos Artemidos. Among the items found were 180,000 mummified cat 'votive' mummies. In February 1890, according to newspaper reports, 19 tons of ballast from a couple of cargo ships were auctioned on the dockside at Liverpool. The ballast was made up of the remains of the 180,000 cats. Some were sold, for museum exhibits or for inclusion in curiosity cabinets and private collections. The ones who were not sold were taken away to be scattered as agricultural fertiliser.

I still have a vague sense of guilt about the one taxidermy specimen I own. I would never buy one but I've watched taxidermied creatures over the years in antique shops becoming more expensive and more dilapidated. They appear with decreasing frequency – old shrews, cross-eyed weasels, red squirrels disfigured by mange. The specimen I didn't buy is a large, pink-toed Peruvian tarantula in a framed glass box. She was bought by a family member and in the processes of time, abandoned and I felt that the least this mighty creature should be given was a place on my study wall. *Avicularia urticans* can live for up to thirty years. I don't how old she was when she died or was killed, or who it was who killed her. I don't know what processes were carried out to preserve her but every time I look into the glass box, I feel uneasy about her. It is not her formidable arachnid presence that concerns me – it is who she was and who she is, the keeping of her corpse and what, in image, idea and 'materiality' she tells me about the boundaries dividing 'them' from 'us'. Hers is the silent commentary of all dead things. In an article about collecting, the writer Audrey Niffenegger describes her sympathy for unloved taxidermied creatures, her collection that includes a toad with 'wires poking out where its hand should have been', a mongoose with a cobra wrapped round it which her boyfriend found 'anti-aphrodisiac', a grey squirrel with an alarmed expression. She writes of all the others she doesn't want to collect and of: 'the spirits of all the unacquired taxidermy standing anxiously at our front gate wondering why they haven't been allowed to come in'.

Despite Thoreau's assertion, I love this museum, the atmosphere, the building, my own enduring, imperceptible place in it. Before I leave, I go back to see the elephant and giraffe, worn now, old and fading. The giraffe must have come from somewhere in Africa and the elephant from Asia. He was probably brought from India to be part of a travelling menagerie. He was made to pull a cart until he was put into the 'Scottish Zoo and Variety Circus' in 1897. In 1900, as a result of normal seasonal hormonal changes, he became aggressive, regarded as a danger and no longer of any use and so he was shot and taken to the taxidermist Charles Kirk whose establishment at 156 Sauchiehall Street, not far away, opened in 1896. Someone gave the elephant the name 'Sir Roger'. 'What right have mortals to parade these thing on their legs again,' Thoreau wrote, 'with their wires, and, when heaven decreed that they shall return to dust again, to return them to sawdust?'

The day has gathered rain while I've been in the museum and now the sky is a deep and satisfying grey. The rain will begin as I'm driving north over the bridge on the way home, a dark sweep over the city as I look back in the driving mirror. I'll think a bit about the notion of 'home', where it is and what it is that can ever take its place. It's what's called 'site fidelity', that instinct in us all to return to the place from which we came, that fragile concept. While I'm walking down the steps, a deep sense of the familiar surrounds me. A flight of pigeons soars and turns over the roof of the Kelvin Hall across the road, glittering in the watery afternoon light. I watch them but can't tell if they're wild or the fortunate, spoiled inhabitants of a devoted pigeon fancier's loft.

7

Tradition

Once, a long time ago, I spent an evening with a friend. We were both reading quietly and pretentiously. By now, I can't remember the exact manifestation of the pretension, but at the time we were moral giants of seventeen, and more than just reading we were deeply engaged in the meaningful, worthy tradition of not going to musicals. Having refused the opportunity to take part in the kibbutz outing to Haifa to watch a performance of *Fiddler on the Roof,* we were revelling in the exceptional nature of our superiority. The absence of people and consequent near-silence gave the evening a unique sense of peace. The only sounds were the singing of crickets and pond frogs. Everyone except the infirm, infants too young to be inducted into sentiment and schlock, and teenage cynics had boarded buses and disappeared off into the night for a theatre thirty miles away. My friend Reuven and I turned down the opportunity to go. We were bemused by the thought of busloads of other Jews, many originally from Germany (including both Reuven's parents), who had survived the years of the Second World War with unimaginable suffering, setting off to watch the performance (with music!) of a piece of theatre which we believed sentimentalized a past way of life of extraordinary deprivation, denial of civil rights, of racial and religious persecution and prolonged, violent discrimination. I'm not sorry I didn't go. I felt uneasy about the prospect of listening to an actor singing a song called 'Tradition' and though I may have adjusted some of my attitudes over time, the evening stays to remind me to be wary of the word and how it's used, of how it's invoked to justify some of the most outrageous, discredited and damaging human behaviour, much of it involved in the destruction of the lives and endangerment of the future of the many species with whom we live on earth.

In the years since then, I've often thought about tradition, what it is and what it does. I've wondered whether we can continue to think of it as the maintenance of continuity with precious aspects of our past, handed down from generation to generation or whether it has to be questioned and subverted, changed as society and ideas change.

'Tradition' isn't fixed. There isn't a single moment that turns a habit, a once-performed ceremony, a way of obtaining or preparing food, of singing, dancing or lighting a candle into a tradition. Some rituals, festivals and celebrations may be truly ancient, a few years old or increasingly something we've done casually for a couple of years or months or once, enjoy, decide to repeat and call a 'tradition'. Our acceptance of a practice on the grounds of its antiquity may rely only on our not knowing its origins. In the introduction to the book *The Invention of Tradition*, Eric Hobsbawm writes of traditions which are 'quite recent in origin and sometimes invented' and uses examples of royal pageantry, which is thought to be ancient but was invented within the past century or so, and the rebuilding of the parliamentary chamber after the Second World War in the same style as the previous one, to illustrate his point. Bruno Latour too writes of invented tradition when in *We Have Never Been Modern* he cites the recent origins of the Scottish kilt. In writing of it as one of the 'great immobile domains', Latour suggests that tradition is part of the representation of a now impossible vision of an earth that might provide 'progress, permanent revolution, modernization, forward flight'.

It's not the concept of tradition itself that is necessarily inimical – there are plenty of venerable practices in every culture, admirable and beautiful ones worthy of profound appreciation for the antiquity of their origins, their cultural significance, grace and meaning. It's the way the term is used and what it's used for that renders it pernicious. There are too many practices which appear still to exist merely because the once acceptable has been extended beyond its reasonable lifetime into the now unacceptable, all of it defended on grounds of 'tradition'. As a concept, word or form, it might not matter if both word and idea held less power but too often 'tradition' transforms the egregious into the venerated by stamping it with a random imprimatur of longevity, wholesomeness and virtue. There is no more potent defence, it seems, than to invoke the word 'tradition'. It is not accidental that it's employed so often to advertise and sell, to control and excuse or that other species are so tightly woven into what are claimed as traditions, represented as faux-antagonists, deserving victims of their vicious fates.

The pursuit of animals for their use as food, medicine or entertainment too often hides under the guise of 'tradition'. Bringing species to the point of extinction for the medicinal or aphrodisiac use of their horns, livers and bones, much of it clandestine and illegal, the vast majority of it completely pharmacologically ineffective, is tradition. Force-feeding geese in order to fatten their livers as an expensive food item is a tradition. Bullfighting is a tradition. Bull-running, rodeos and hunting particular species at particular times of the year are traditions. 'Tradition' claims quasi-historic rights for people who illegally dig out badgers, chase hares, force dogs to fight, as well as many of the legal human activities which – unaccountably – involve the persecution, the killing and infliction of suffering on large numbers of other, non-human animals. Numberless varieties of animal torture and slaughter all over the world have long been designated 'traditions'.

*

There's a place I go to on a promontory above the city where I can look out over an expanse of sea where cetaceans pass below, white-beaked dolphins, minke whales, harbour porpoises. I watch them, knowing that as I do, I join the many others worldwide who are thrilled and moved by the closeness of their presence and the mysteries of their lives. It wasn't very long after the evening I spent with my friend when the slogan 'Save the Whales' began to appear and spread. The words rapidly became a herald for a new way of looking at the relationship between humans and other species, an expression of profound concern at the prospect of whale extinctions. Suddenly, it was everywhere. In every office, in every student room there would be a Save the Whales mug, poster, sticker, badge, all drawing our attention to the growing campaign against the practice of whaling.

After the depredations of three hundred years or so of 'commercial' whaling, the depletion of cetacean species had begun to become clear in the early years of the twentieth century with the increasing industrialization of the methods of whaling. In 1925, the League of Nations became involved in the question of whaling and over subsequent decades organizations were set up to try to regulate the activities of whaling nations until 1946 when the International Whaling Commission was formed to control and monitor whaling activities although, having no statutory powers, its ability to prevent either commercial or 'scientific' whaling has always been limited. In 1961, 66,000 whales were slaughtered, the highest annual number ever recorded. It would be the growing involvement during the 1970s of organizations such as Greenpeace that would eventually bring about the 1986 moratorium on commercial whaling although, despite the ban, whaling in countries which were not signatories to the treaty continued.

In 1970, Judy Collins sang the song 'Farewell to Tarwathie' on her album *Whales & Nightingales*. The song, originally written in Scots, was composed in the 1850s by an Aberdeenshire miller to draw attention to both the dangers and lucrative rewards of whaling, but when

Collins sang it, it was with a background of whale sounds, recorded and given to her by the biologist Roger Payne who described them as 'the most provocative, most beautiful sounds made by any animal on earth'. Collins donated some of the royalties of the hugely successful album to Payne, founder of the organization Ocean Alliance. The moment was significant – the song became an anthem for the Save the Whales campaign at a time when, although it had been known for decades that whaling was destroying the life of the seas, people were becoming newly aware of the damage humans were doing to the natural world. In the past century alone, whale numbers have reduced from 5 million to 1.5 million.

In September 2018, at a meeting of the International Whaling Commission in Brazil, the 'Florianopolis Declaration' was passed, declaring whaling no longer a 'necessary economic activity'. The non-binding declaration was agreed by forty countries and rejected by the twenty-seven who support whaling, the most important of whom is Japan. The IWC's moratorium on whaling isn't recognized by Norway, recognized only in limited form by Iceland and has exemptions for 'aboriginal subsistence' and for 'scientific' whaling, the dubious rationale used by Japan for its continuation. Both Iceland and Norway set their own arbitrary quotas, widely regarded as being unsustainable. Iceland, under the aegis of the single company which does it, allows the hunting of endangered fin whales. Greenland is given an 'Aboriginal Subsistence Whaling' quota by the IWC.

Every account of the slaughter of large cetaceans whether factual or literary describes prolonged, extended suffering during the process. The North Atlantic Marine Mammal Commission, an organization set up by the northern whaling territories (Iceland, Greenland, the Faroes and Norway) produces a set of manuals on how to carry it out and the weaponry to use: Kongsberg and Henriksen whaling guns with harpoon and explosive whale grenades, large calibre rifles, blowhole hooks and spinal lances. Among the 'small cetaceans' whose killing they discuss, are harbour porpoises, white-sided and white-beaked dolphin, killer whales, narwhals, beluga and pilot whales.

Whaling may be cruel, economically unnecessary and environmentally damaging but the stated justifications for it given by the main territories that still carry it out – Iceland, Norway, Japan and the Faroes – are all much the same. It's tradition. In a paper on Japan's response to international pressure to stop whaling, Chris Burgess, a lecturer at Tsuda Juku University in Tokyo, suggests that rather than being a matter of environmental concern, the debate is about national identity and pride, likely only to be strengthened by continuing criticism and opposition.

The provision of whale meat as a traditional food is often used as justification for whaling. In the Faroe Islands, a tradition called the *grindadráp* persists. Regarded by many as a vital aspect of Faroese social cohesion, it is seen as a method of providing food for a marginal community. When pods of pilot whales approaching the Faroese coast are sighted, often in their hundreds, they're driven from boats into the shore to be killed by islanders who, with spinal lances and hooks, butcher the creatures en masse, *en famille*, turning the sea to blood. Pilot whales, long-lived, sociable, moving in a stable matrilineal group are the species most likely to become beached. One Faroese official describes the *grindadráp* as 'ecological and respectful ...' while a Faroese government website states: 'Local and traditional forms of agriculture and hunting, including coastal drive hunts of pilot whales have enabled the Faroe Islands to maintain a relatively high degree of self-sufficiency in food production.'

In 2016, the founder of the Faroese metal band Týr, Heri Joensen, wrote in defence of the *grindadráp*, suggesting correctly that we hold different attitudes towards wild creatures from those we slaughter in slaughterhouses. His reward for his participation in the killing, he wrote, was 150lbs of valuable whale meat which, in most circumstances, might seem like something of a deal if there weren't the strongest medical recommendation that no one should eat whale meat at all. Heavily polluted by mercury, toxaphene, chlordane, DDT and PCBs, the meat and blubber of cetaceans, the top of the marine food chain, are peculiarly susceptible to the effects of ingested organochlorines. For years, the leading Faroese researcher Dr Pál Weihe has studied the effects of eating whale meat – among them, increases in the risk of developing Parkinson's disease, hypertension, arteriosclerosis of the carotid arteries and cognitive and immune system impairments in children. The recommendation that only one meal of whale meat should be eaten in a month was recently revised to suggest that none should be eaten at all. There are those who are suspicious of the advice, believing that it's a ruse to stop people enjoying the freedom to carry out the *grindadráp*.

In other whaling countries too, few people choose to eat whale meat, either because it's not particularly palatable or because they are sensibly concerned about its toxicity. Despite extensive government subsidization of whaling and efforts to encourage the eating of whale meat, it's unpopular in Norway where stocks have built up in cold storage facilities, as they have in Japan. In 2015, Japan rejected imported Norwegian whale meat because of its excessive pesticide levels. To provide justification for the continuation of whaling, Iceland encourages tourists to eat it on the grounds that it's 'traditional', a claim which is disputed. Japanese government limits for mercury in seafood are four times the levels allowed in the United States but don't apply to cetacean meat, which studies have shown may contain levels of mercury contamination up to 5,000 times over safe limits for consumption. Greenland, despite its quota for 'aboriginal subsistence', encourages the eating of whale meat by tourists. An international study carried out at Aarhus University in Denmark published in 2018 details of PCB and other pollutant levels in whale populations and likely effects on future cetacean breeding success. Predictions are that as a result of widespread pollution, some cetacean species may disappear completely within a few decades. In a letter to the *Financial Times* in 2018, Chris Butler-Stroud, CEO of Whale and Dolphin Conservation, described how whaling activity has removed 80 per cent of whale biomass from the oceans, and that the last 20 per cent is being destroyed by 'ocean acidification, fisheries, by-catch and pollution'.

In a paper 'Whales as Marine Ecosystem Engineers', published in 2014, the authors discussed the vital, complex role large cetaceans play in marine – and thus all – natural ecosystems by their stirring up of nutrients, their contribution to feeding cycles by physical means through faeces, urine and placentas, their part in 'trophic cascades' – the interconnected and complex ways in which top predators influence the food chain – and in death, adding their bodies to the organic enrichment of the deep seas. The presence of whales stimulates the growth of populations of phytoplankton, which not only contribute 50 per cent of all oxygen to the atmosphere but capture large quantities of CO_2 – around 40 per cent of all the CO_2 produced. Recent research has demonstrated that allowing whale populations to increase to their pre-whaling numbers would have a significant effect on ameliorating climate change through carbon capture by the whales, as well as by increasing production of phytoplankton. The possible effects of increasing whale numbers have been estimated by the researchers as equivalent to the amount of CO_2 that would be captured by the planting of four Amazon forests.

Empathy for whales, as exemplified by the Save the Whales campaigns, seemed a new phenomenon, an expression of concern about the affairs of humans no less than those of animals. For the first time, many people identified the deleterious effects of human activity on earth as a matter of political determination. Herbert Marcuse, who spoke eloquently about the destruction of the natural world, described the war in Vietnam with its deliberate destruction of the environment as 'ecocide'. The words 'Save the Whales', effective though they were in raising the issue of the slaughter and bringing about important political action, became one of the ways environmentalists and those concerned with the growing threat to other species were mocked as hot-headed, hippies, the oversensitive young whose regrettable propensities for caring would no doubt pass, given sufficient time. It all seems so long ago now.

The human relationship with whales was never simply one of exploitation and death. Cetaceans have flitted through our centuries, through the depths of our thoughts, imaginations, literature and history, intriguing and impressing us by their scale, their size, by their all too evident intelligence, their habits, social and family organization, their capacities for empathy and moral behaviour, the impenetrable, unknowable aspects of their lives and often brutal nature of their deaths.

Since the discovery of the sounds made by whales and Roger Payne's subsequent production of the album *Songs of the Humpback Whale*, their stirring musicality has elicited widespread human interest. In 1980, Greenpeace commissioned the composer Toru Takemitsu's *Toward the Sea* for the Save the Whales campaign, part of a body of whale-inspired music which includes John Cage's *Litany for the Whale* and the contemporary composer Emily Doolittle's *Social Sounds from Whales at Night*. In his fascinating study of whale music *Thousand Mile Song*, the musician David Rothenberg quotes a fellow musician describing whale sounds: 'They had this bluesy quality that was so poignant. . . . there is perhaps a universal yearning that is shared by all species, this calling, crying quality, in their singing.' Rothenberg describes whales as 'subterranean jazz musicians'.

In his essay 'A Presentation of Whales', Barry Lopez gives an account of the stranding of forty-eight sperm whales on a beach in Oregon. The whales, deep-sea creatures who ranged in size from thirty to thirty-eight feet, became beached one June evening in 1979. The incident caused unprecedented confusion and uncertainty, and was accompanied by displays of what Lopez described as 'the best and worst of human behaviour'. In the essay, Lopez presents some of the profound questions around our relationship to other species and the extent of our empathy, or lack of it. No one knows why whales strand. In *Historia Animalium*, Aristotle wrote about the phenomenon: 'It is not known for what reason they run themselves aground on dry land . . .' In Oregon, as happens in most cases of whale stranding, it was unclear what should be done. Could the whales be refloated? Could they be saved? Should they be? How long should they be left to suffer?

The ensuing days of scientific, veterinary, police, public and press involvement were chaotic although fairly soon the outcome became clear – as soon as the whales beached, they were certain to die. People scrambled over dying whales to take photos and tried to remove teeth for 'scrimshaw' while scientists struggled to find ways to alleviate the whales' suffering. Lopez described an incident when the blubber was being removed from a dead whale and the one next to her pounded the sand with his flukes, continuing for fifteen minutes. He wrote of the whales' 'purple brown eyes', of their sounds, their 'mewing and clicking' and of the terrible task of dealing with the aftermath, the cutting and the burning of the huge creatures, the immense sadness that surrounded these mysterious and terrible deaths.

In his memoir *Island Home*, the Australian writer Tim Winton describes growing up in Albany, the last of Australia's whaling ports and his realization as a teenager that the sperm and humpback whales who were being processed at Albany were facing extinction: 'All this carnage and waste for fertilisers and cosmetics. What had always seemed normal was suddenly as absurd as it was grotesque.' His awareness of the power of protest and sense of the urgent need for protection of the environment made him act towards preserving and restoring the natural world.

The whaling station at Albany lies abandoned now, one of a worldwide network of the remnants of the whale-killing industry, extending from the South Atlantic in South Georgia's Stromness Bay, Leith

Harbour, Husvik and Grytviken, to South Africa at Durban and Donkergat, in Spitsbergen, Perano in New Zealand, Bunabhainneader on Harris in the Outer Hebrides and many more; these eerie, empty monuments to death and greed, the places where, over centuries, creatures were laid bare, 'flensed', reduced to oil and blubber, almost extirpated from the seas.

If the rationale for whaling, that of providing meat for food has gone or is seriously compromised, why continue doing it? Because, however meaningless an activity, however damaging to both human and cetacean, however futile and bloody, however much it displays little more than stubborn contrariness and a possibly uncontrollable urge to kill, it's tradition.

Tradition. Christmas, Easter, the High Days and Holy Days, Thanksgiving, the days and occasions when we slaughter in the name of tradition. It is annual, biennial, seasonal, our years marked by these atavistic depravities. Everywhere, in every place, on land, in forests, mountains, plains, in battery units and factory farms, on islands, in open sea and skies above, one or other species is marked as prey, bred for the purpose, hunted down, selected, designated by 'tradition' for slaughter. Perceptions of religious observation often blur the boundaries between doctrine and 'tradition' as practices more cultural or seasonal than religious become identified as integral to religious law.

Among some ultra-orthodox Jewish communities, an annual practice known as *kapparot* – 'atonements' is carried out. It takes place before Yom Kippur, the Day of Atonement, as part of the season during which the observant are meant to consider their behaviour of the past year and, if necessary, make amends. Kapparot involves seizing an unfortunate chicken who is then swung three times around the head in symbolic transference of sin, before the chicken is slaughtered and donated to charity. (I've never seen it done and didn't even know about it for a long time – the stolid, middle-class Scottish Jewish community into which I was born would never have participated in or condoned chicken torture.) It has long been the subject of dispute and the kind of fine and detailed discussion so beloved of scholars of doctrinal minutiae. With comparatively recent roots in early medieval custom, kapparot has been both defended and opposed by religious authorities ever since, although the case for the defence is considerably more opaque, arcane and tortuous than the case for the opposition. Condemned by the sixteenth-century codifier of Jewish law Joseph Caro as 'a foolish custom', it was defended by his contemporary Rabbi Moshe Isserles on the deeply dubious grounds that the longevity of the practice justified its continuation. As ever, the contradiction of kapparot lies in the Jewish laws concerning the proper treatment of animals, *tsa'ar ba'alei chaim*. If you accept that God told you to treat animals kindly, to feed them before you feed yourself and to avoid cruelty, then you do not, if you're a vigilant adherent to the teachings of your faith, whirl chickens round your head, keep them in crates and all the other horrors to which these birds are subjected every Yom Kippur. If you're decent, humane and truly devout, you don't perform kapparot or defend it by calling it either religion or tradition.

The cookery writer Elizabeth David's book *Italian Food* was first published in Britain in 1954. In it she describes the ubiquity of bird-eating in Italy and the indiscriminate killing of birds of every sort and size, 'for the fun of the chase and for the benefits of the table'. These included: 'thrushes, larks, robins, blackbirds, bullfinches as well as quails, ortolans, snipe, woodpeckers, figpeckers – in fact almost any-thing which flies.' Flamingoes, she said, were eaten too, as one assumes were the spotted eagle and three storks her friends saw for sale outside a butcher's shop. David's recipes include those for *Beccafichi al Nido* and *Ucceletti* – 'Figpeckers in their Nests' and 'Small Birds'.

In the years since the book's publication, despite growing awareness of species depletions and the introduction of effective EU laws such as the Birds Directive to restrict and control hunting, particularly during migration, the shooting and trapping of small birds in vast numbers has continued, all of it carried out in the name of 'tradition'. In certain places in particular seasons people dress in military fatigues, take up assault weapons, set up mist nets and traps, get out their lures, their baits, their electronic calling devices and their limesticks and begin the lucrative, traditional business of killing birds. The numbers of migrating birds still slaughtered every year in Malta, Cyprus, Italy and the countries of the Middle East, both legally and illegally, are estimated to be up to 36 million. These birds belong to some of the 457 or so species occurring regularly in the region, among them blackcaps, common quail, chaffinches, house sparrows, song thrushes, European turtle doves, curlews, ferruginous ducks and rock partridge. The latter three are killed in substantial numbers although critically endangered. Huge numbers of migrating swallows, swifts and bee-eaters are slaughtered during migration in the skies over Lebanon, robins are trapped in their thousands in Italy and entire flocks of white storks are killed because the impulses which direct their migration routes takes them over Malta.

In one of the most moving and horrifying accounts of these practices, the American novelist and appreciator of birds Jonathan Franzen wrote in the *New Yorker* in 2010 of the bird killing in Cyprus, Malta and Italy, about its extent, its political and black-market connections, its exceptional cruelty. Describing the use of limesticks he wrote: 'Stuck in lime in the second orchard were five collared flycatchers, a house sparrow and a spotted flycatcher ... as well as three more blackcaps ...' He wrote of finding a thrush nightingale, so badly injured that he or she had to be killed. He described the acacia groves, the mist nets, the lures of the trapping sites, the superstition among poachers that to let unwanted birds go free ruins a trapping site and so 'the unmarketable species are torn up and dropped on the ground or left to die'. He described Malta as 'the most savagely bird-hostile place in Europe' with the unbridled shooting of 'bee-eaters, hoopoes, golden orioles, shearwaters, storks and herons ... swallows ... migrating hawks ... endangered raptors, such as lesser spotted eagles and pallid harriers, that governments farther north in Europe are spending millions of Euros to conserve.'

In the time since he wrote it, this Mediterranean enthusiasm for bird killing is unabated. 'Tradition' may be a well-worn defence for the activity, but money is a primary motivating factor too. Mass bird slaughter is underpinned by serious criminal activity, the involvement of organizations from which most people would rather maintain distance, the encouragement of the arms trade and a severe lack of local political will, all nestling safely under the brittle defence of 'tradition'. It takes considerable courage to do as the dedicated members of bird organizations, journalists and assorted naturalists have done over the years to confront and monitor the activities of armed, organized and violent 'hunters'.

Cyprus is another place where the wholesale trapping and killing of birds takes place. One target species is the blackcap, *Silvia Atricipilla*, particularly favoured for inclusion in a dish called 'ambelopoulia', a 'traditional delicacy' composed of whole small, roasted birds. Although banned since 1970, this dish seems both clandestine treat and gesture of defiance against the EU laws, which might control the activity. Reporting a demonstration against the threat of restrictions to 'the tradition of bird trapping', the *Cyprus Mail* of October 2017 included reference to a group which calls itself 'the friends of limesticks'. In the same year, objecting to a possible ban on bird-trapping, Archimandrite Avgoustinos Karras of the Bishopric of Constantia said of bird lime: 'This is tradition, not a crime' seeming not to have considered that it may be both.

Blackcaps come to my garden to feed at the bird table, lovely birds of muted greys with the high, bright, complex calls that have given them the name of 'northern nightingales'. They're increasingly common in Britain now, encouraged by climatic changes and our provision of food. I watch them as they feed. They seem discerning in their habits, in their ability to know, inexplicably, which food will be on offer, appearing only when I put out nyjer seed. How do they know? What unfathomable system of scent, observation or communication tells them what's here or not?

On her list of edible species, Elizabeth David also included the ortolan bunting, *Emberiza hortulana*, a bird whose population has declined by over 80 per cent since the 1970s as a result of environmental factors such as climate changes and habitat loss, and of being hunted during migration. Migration is a difficult, harrowing prospect for anyone, human or not. For birds – so small in relation to the extended journeys they have to undertake, crossing continents and oceans – it is particularly hazardous, costly in the energy they require to begin the journey, and the dangers they face en route, from the vagaries of weather to predation and accident. Despite a ban on hunting ortolans since 1979 and their being made a protected species in 1999, until recently 30,000 were illegally trapped and killed annually, particularly in south-west France where they were (and almost certainly still are although in smaller numbers) sold to the restaurant trade. In preparation for being consumed by wealthy diners, ortolans are blinded to disrupt their feeding habits, kept in tiny cages and force-fed before being drowned in Armagnac. The diners, by tradition, cover their faces with napkins as they eat. In his book *Medium Raw*, the late chef Anthony Bourdain wrote of eating smuggled ortolans in New York; his description of skull-crunching and organ swallowing is as grotesque and repulsive for its self-indulgence as it is for its detachment from any sense of responsibility or empathy. It is estimated that northern populations of ortolans, the very ones who fly over southern France, are 'directly threatened with extinction'.

Extinction. For a long time, the concept of extinction wasn't obvious, at least not to the many in the Western world who, up until the mid-eighteenth century, pondered questions of creation. Anyone disinclined to contemplate anything but the Christian view of the origins of life wouldn't easily accept (or dare to suggest) that God might have been as ill-considered as to create species only to allow them to perish and disappear in their entirety. The discovery of fossils provided opportunity for imaginative explanations for their presence, from Aristotle's belief that they were sea creatures who had swum into cracks of rock and become stuck, to those who believed that they were works of the devil or visitations upon the evil-doer responsible for bringing about the Flood. The first written record that displays an understanding of extinction was made by Xanthus of Sardis in 500 BCE in his observation that creatures embedded in rock were the remains of creatures once living. Leonardo da Vinci, a dedicated and perceptive geologist in addition to the many other manifestations of his extraordinary genius, was deeply sceptical of the ideas put forward to explain the presence of fossil shells among the rocks of northern Italy, among them that they had been deposited after the biblical flood or that they had simply grown there. Appreciating that fossils were creatures once living that had been laid down over the course of time by the action of seas, he wrote, 'sufficient for us is the testimony of things created in the salt water, and found again in the high mountains far from the seas'.

Robert Hooke, brilliant natural scientist, architect, engineer, astronomer and writer of the first book on microscopy *Micrographia*, published in 1665 and illustrated by his exquisite drawings of fleas, flies, gnats and other 'Minute Bodies', appreciated that fossils were the remains of life forms now died out, putting forward the argument to religious critics that since all species would come to the same at the end of the world, God was demonstrating the extent of his providence by killing them off sequentially.

Although others had approached the idea, it was Georges Cuvier, appointed as professor of anatomy at the National Museum of Natural History in Paris in 1795, who was the first to provide palaeontological evidence of animal extinctions. In a lecture he gave to the National Institute for Sciences and Arts the year after his appointment, he illustrated the differences between the bones of animals still living and those of large fossil animals, suggesting that since there was no evidence of the latter still being found alive, they were clearly now extinct. He later demonstrated that mastodons, Irish elk and giant ground sloths, to which he gave the name *Megatherium americanum*, had joined others in their woeful fate.

Today, when there's so much more awareness of species extinction and its implications, it might be possible to believe that it is a Holocene thing, an Anthropocene thing, but it's not – extinctions have always happened as species have been lost in vast numbers during the course of evolution as they've died out naturally or from extrinsic causes, during the five great mass extinction events or as a result of the natural processes of 'background extinctions', those attributable to climactic, atmospheric and environmental factors. Background extinction rates are calculated according to a formula of E/MSY, or extinction per million species per year. The 'natural' background rate has been estimated from fossil records at between 1 and 0.1 E/MSY. Current extinction rates are calculated to be 1,000 higher than the natural background rate and, it has been suggested, are likely to increase 10,000 times more in the future. It's the speed of decrease in species numbers that's the singular phenomenon of the Anthropocene extinctions and the cause of our arriving so catastrophically at the point where the current state of the earth's biota is described by many as 'the sixth extinction' or 'biological annihilation'. In a paper in 2017, Gerardo Caballos and other authors wrote: 'Humanity will eventually pay a very high price for the decimation of the only assemblage of life we know of in the universe.' Threatened with extinction are 26,000 species. Humans are, one scientist says, 'cutting down the tree of life'.

During the aftermath of the terrible An Lushan Rebellion in eighth-century China, the poet Po Chu-yi wrote of the migrant goose he had rescued from a market. Setting the goose free, the poet exhorts him to take care to avoid the starving bands of soldiers who will try to catch and eat him. In 2010, the American poet W. S. Merwin wrote a poem, 'Message to Po Chu-yi', reporting that the goose was well and with him and had been for a long time. Informing Po Chu-yi of how times have changed, Merwin wrote of contemporary life:

> . . . I will not tell you what
> is done to geese before they kill them
> now we are melting the very poles
> of the earth but I have never known
> where he would go after he leaves me

Extinction seems a concept far from the hunter's slogan, the words 'If it flies, it dies . . .' written on T-shirts and used as the words of a song. On European flyways and migration routes, every passing species of birds is shot, some in high numbers and traded illegally across borders: protected species of pipits, Eurasian skylarks, warblers, shot in Romania and shipped via Slovenia for the clandestine restaurant trade in Italy. In June 2018, finch trapping was outlawed in Malta to protect siskins, goldfinches, European serins, linnets, greenfinches, chaffinches and hawfinches. The trapping of song thrushes and golden plover is, under an EU derogation, still legal. In Finland the hunting of eider is allowed despite its continuation being referred to the European Court of Justice because eider populations are 'not at a satisfactory level'. According to the EU Commission's finding: 'In December 2016, the Commission sent a reasoned opinion, urging Finland to stop the spring hunting of eider males in Åland Province. However, the competent authorities decided to open a new spring hunting season in April 2017.'

In Åland, a local hunter of eider describes taking his young son with him to teach him how to hunt. He wants to carry on because: 'it's the way we live here'. It is, we are to believe, all about taking your son out killing with you, teaching him to kill, sitting round talking about killing, looking forward to a similar lifetime of seasonal killing.

Questions of further endangering bird numbers seem far away from these tooth-piercingly saccharine defences of the selfish exploitation this hunting involves. Sentimentality seems a poor and insubstantial excuse for causing species extinction.

Turtle doves are red-listed as 'vulnerable' by the International Union for the Conservation of Species, their population in Western Europe nearing extinction. Malta's government still allows them to be hunted. 'It's part of our life, passed from one generation to another. At the end of the season, we sit and cook together. There are songs about turtle dove hunting in spring . . .' a Maltese hunting leader declares in a tone of bewildered, defensive petulance. There may well be but there are songs about suitcases too, about shoes, revenge, murder, serial killers, lynchings, migrant workers dying in plane crashes and royal executions. There are songs about trees and mice and rivers, about naughty sprites who impregnate passing maidens and befuddled men who shoot their girlfriends mistaking them for ravens.

In 1934, the biologist Sir Julian Huxley directed a short film, *The Private Life of the Gannets*. Made on Grassholm island off the Pembrokeshire coast of Wales, now an RSPB reserve, it was produced by Alexander Korda and was the first wildlife film ever to win an Academy Award. Remarkable for its detail, the close-ups of gannet eyes, of an egg hatching, of courtship, diving and nesting, it was accomplished without the sophisticated equipment that allows us the intimate observations and almost-intrusive clarity of current wildlife documentaries. Accompanied by sympathetic commentary on the habits and behaviour of these spectacularly beautiful creatures and a suitably romantic score, it is a unique portrait of time, bird and place. The year after it was made, Huxley wrote in the magazine of the Royal Geographical Society, questioning the continuation of one particular, long-standing Hebridean practice involving gannets, but despite his intervention and the debate in the decades since, the practice continues, surrounded by every known layer of social, legal and religious protection, stoutly defended under the name of 'tradition'.

Every summer in Scotland at some secret, guarded moment on an early, possibly glowing, possibly loomingly grey, cold and pouring morning in the northern Hebrides, a group of ten men will set out to sail from Ness at the northern point of the Isle of Lewis, the northern-most of the Outer Hebrides, to Sulasgeir, an isolated isle of rock, 40 miles out in the Atlantic, as they've done every August for centuries. There they will, with considerable difficulty and discomfort, set up camp among steep rocks and in old bothies to begin the preparation for their self-appointed task of spending the following fortnight seizing from their nests, killing, eviscerating, plucking, salting and stacking 2,000 young gannets who will then be taken back to Lewis to be sold and eaten, or sent to Lewis people living elsewhere. The men, more often ten but sometimes twelve, call themselves 'The Men of Ness' and their endeavour is called 'the guga hunt'. 'Guga' is the Gaelic name for the young of the northern gannet, *Morus bassanus*, or Solan goose, a large seabird of the Order *Suliformes*. Despite seabirds being protected 'at all times' under the Wildlife and Countryside Act 1981 and the Nature Conservation (Scotland) Act 2004, special exception is made to allow it. The 1981 Act prohibits the use of particular items for catching or killing birds. Crossbows are prohibited, the use of chemical wetting agents or live decoy birds if they are tethered, blinded or maimed. Traps such as gins, snares and springes – a type of spring-loaded noose – are prohibited.

The origins of the killing of guga lie in necessity – life in the islands has always been formidably harsh – but it is the antiquity of this voyage, first recorded in 1549, although certainly taking place long before, which encourages some to see this annual expedition as reso-lute homage to cultural continuity, a brave flare of defiance in the face of the erosive incursions of the modern age, or as a challenge to the incipient dangers of modern, or postmodern, secular thought.

A great deal that surrounds the 'guga hunt' encourages the viewing of it through a heavy atmosphere of hushed religiosity. A long historic line connects an early Christian presence on both Sulasgeir and its twin island of North Rona to the stern, creationist Christianity of the Free Church of Scotland, which is central to the lives of the people of Lewis. The early eighth-century monk St Ronan is said to have lived on the greener, gentler of the two islands, North Rona, where the ruins of his chapel, cell and oratory stand in wind-blown endurance. Ronan's sister, the unfortunate St Brenhilda, apparently in response to her brother's decidedly un-monkishly incestuous advances, is believed to have fled alone to the lonely, bird-inhabited island. In his poem 'St Brenhilda on Sula Sgeir', the Irish poet David Wheatley described her eremitic endurance:

> . . . living on guga
> and cress, telling my prayers
> by the light of a cormorant lamp,
> its Pentecostal tongue . . .

The finding of Brenhilda's skeleton, a seabird nesting in her ribcage, has long provided a haunting, premonitory metaphor for the disparate powers of man, woman and nature.

In his book *The Old Ways* Robert Macfarlane described the sighting of the two islands from the south as feeling 'as if you have sailed into a parable'. He wrote too of the lengthy preparations that take place on the first days the 'Men of Ness' are on Sulasgeir, the transportation of supplies, the equipment for securing the boat, for providing shelter in the stone bothies, for constructing the pulley and chute system to facilitate the movement of goods up and down the rock face, the designated tasks of the protagonists as they set about killing, plucking and de-winging their prey.

The sea journey from Ness to Sulasgeir itself has been widely, fulsomely described by both observers and participants in word, photo, film and song, often in the mistily romantic terms used for travels on the western edge of Europe, in these places of quietude and beauty, the last island landfalls before America where you can stand at sunset on high promontories to gaze over wide ellipses of still and gilded sea, believing for a moment that they extend only as far as the horizon, the sharp rim at the end of the world, the end of your line of sight.

In 1697, Martin Martin, a native of Skye, undertook a similarly hazardous sea journey of which he gave account in his book *A Late Voyage to St Kilda*. He wrote of his close and sympathetic observation of gannets, using their older name 'Solan goose':

> The *Solan* Geese have always some of their Number keeping Centry in the Night, and if they are surprised, as it often happens, all the Flock are taken one after another; but if the Centinel be awake at the Approach of the creeping Fowlers and hear a Noise, it cries softly, *Grog Grog*, at which the Flock move not; but if the Centinel sees or hears the Fowler approaching, he cries quickly *Bir, Bir* which should seem to import Danger, since immediately after, the whole Tribe take Wing, leaving the Fowler alone on the Rock to return home *re infecta*, all his Labour for that Night spent in vain.

Large and beautiful birds, gannets are brilliant white, wide of wing, all points and sharpness, all bill and wings and shaped for speed and high, precision diving. Their wings are black-tipped, their faces tinted palest yellow, their eyes the blue-ringed, dark-lined eyes of Ptolomaic queens. They form lasting pair bonds, return annually to the same nesting sites and gather in huge colonies, 'gannetries', on the coasts and outlying islands of Britain, at Grassholm Island in Wales, Bempton Cliffs in England, round the Scottish coast on the Bass Rock, St Kilda in the Outer Hebrides and Sulasgeir. They're careful, devoted parents, noisy, aggressive neighbours.

The physiology of gannets is unique in the precise shape of their thickened skulls, in their long necks – more than 50 per cent of their body length – and in their slender, elongated bones. In a research study published in the Proceedings of the National Academy of Sciences in 2016 entitled 'How seabirds plunge-dive without injuries', Sunny Jung, a biomedical engineering and mechanics professor at Virginia Tech, and his colleagues, examined the mechanisms by which seabirds such as gannets are able to dive from height and at speed without sustaining the organ, bone and tissue damage which would be experienced by human divers. Jung describes the diving of gannets as being like 'torpedoes hitting the water'. Using 3D-printed replica bird skulls from the Smithsonian Museum, CT scans of a dead bird at the local veterinary hospital, and a salvaged gannet (frozen to maintain the required rigidity for repeatedly dropping the bird into a tank of water from considerable height) they were able to pinpoint the moment of peak danger to the bird during the process of diving. The researchers isolated maximum vulnerability to the point at which the bird's head only is immersed, making it susceptible to the drag force from the water. They discovered that the relationship between the bird's impact speed, the length and construction of its neck and the shape of its head and beak all contribute towards preventing the bird's neck buckling at the point of impact. With this research, Jung and his team have been able to ascertain safety limits for human diving. Iain Stenhouse, director of the Marine Bird Program at the Biodiversity Research Institute in Portland, Maine, suggests that gannets are aided in their diving by internal structures resembling airbags, and by their particularly resilient skulls.

I hold part of a bird's skull in my hand. It belonged to a seabird but I don't know which. It's too small to be a gannet's and the beak's missing. I no longer remember where I picked it up, on which beach, how long ago. The smooth bone's dried out now into the grey-white fineness of Japanese raku, this once-container of a vigorous avian brain.

Some of the other skulls with whom it shares the mantelpiece seem less fragile, the rook's, the rabbit's, the pigeon's. To smash it would take less effort than the smashing of the skull of a young and living bird.

In response to a letter of concern from protestors in 2010, the Scottish government wrote:

> It is (our) understanding that most of the gugas will be killed by a single blow to the head. Where a second blow is required, it is very likely that the first will have rendered the bird unconscious. In our view therefore the method used to kill the gugas does not involve unnecessary suffering. Given the above and that the guga hunt is carried out in accordance with a licence issued by the Scottish Government we are confident that the guga hunt is compatible with the requirements of section 19 of the Animal Health and Welfare (Scotland) Act 2006, which allows the killing of an animal in an appropriate and humane manner.

Unnecessary suffering. What is necessary suffering? The law allows the annual killing of 2,000 birds although an expert on gannets tells me that many more birds probably die or are killed as a result of the 'hunt' because of the disturbance to the colonies while it's taking place. The results of one report suggest that the killing of gannet chicks (referred to in the report as a 'harvest') reduces the rate of population growth that would occur on Sulasgeir if the activity did not take place. Gannet numbers have increased at colonies in Britain and Ireland in the past forty years and new colonies have been established, but this increase has not been replicated at Sulasgeir where there are unused nesting sites and where it is probable that population numbers rely on immigration from other colonies. The question though is not about numbers, it's about cruelty.

In 2011, a film of the hunt made by the film-maker Mike Day was televised. In a later account of the difficulties involved in the making of it, Day reported that among the stipulations he had to agree to before being given permission was that the killing of the birds should not be filmed. Since few people outside the few chosen and designated 'hunters' are allowed to observe it, there is little way of knowing if the killing of the young birds is accomplished as the participants claim, with expeditious speed and as the government claims, without 'unnecessary suffering'. The words used to describe the killing, even in sympathetic accounts, are contradictory: 'painless', 'clubbing', 'instantaneous', 'beating'.

In his book *The Guga Hunters*, the most comprehensive on the subject, Ness native Donald Murray provides descriptions of the most desirable birds for catching and killing: age, size, colour, details of the implement used for catching the young gannets too: 'a long pole with a spring-loaded jaw on top', and how the bird is lifted from the nest with this device by one man and passed to another who kills it with 'one blow from a heavy stick'.

The 'guga hunt' has been described by participants and defenders as 'legendary', 'heroic', 'traditional'. It has been argued that the people who carry it out are 'natural environmentalists' (although travelling to an island to kill birds does not make you an environmentalist. It makes you a person who travels to an island to kill birds.) The language and imagery of war are frequently invoked in its defence. Donald Murray ascribes moral justification for the hunt to the bravery that sustained many past participants through their roles in two wars. The photographer of a book of black and white photographs of the hunt says that Sulasgeir reminded him of 'Passchendaele and Calvary'. The spiritual is invoked too in references that permeate the narratives like a fine structure of invisible steel. One Hebridean writer on ecology quoted by Murray believes that for a young gannet to be killed in this way is morally defensible if the killing is 'providentially grounded . . . replete with gratitude for the grace of God . . .' an argument not infrequently used by those who start wars, who conquer and occupy other people's land believing their deeds not only acceptable but necessary and sanctified because God whispered something secret into their ever-receptive ear. No deed may reflect back on the righteous man.

I phone an organization involved in the protection of Scottish wildlife to ask about their view of the 'guga hunt'. The woman who answers the phone hesitates. 'It's very – political,' she says. I tell her that I know. She doesn't want to discuss it but says that since it's legal that's just the way it is and hangs up. I phone other organizations, am given numbers, and more numbers but there are no replies. I leave messages. I try again. More hesitation, then a man is found who'll speak to me. He's senior in the particular organization and angry that I'm asking and aggressively tells me a lot of things I already know and that it's tradition and continues as a result of the tragic dispersion of Scottish Highland and island communities and the subsequent destruction of traditional ways of life. Do I understand? I do but don't tell him how or why. I don't say to him that I come from a tradition of traditions, from a culture of memory and longing, from a long and stratified exilic past and the memory of the destruction of traditional ways of life. It is this, I want to say to him, which has made me mistrustful of 'tradition', of accepting too easily, of continuing because one continues, of not asking why we do.

Guga is a 'traditional dish' described (not widely) as 'a delicacy'. A Hebridean friend I ask pretends to gag at the memory. Methods for its preparation are unusual. Donald Murray talks of the birds being washed in soda crystals or in washing-up liquid before being cooked. One blog that deals lightly with the intricacies of Hebridean cooking provides instructions for marinating the bird in light diesel oil (or in the absence of diesel oil, tractor paraffin) for a year before cooking it. A further post suggests that a more modern approach is to cover the bird with a well-known industrial hand-cleaning gel before roasting.

There is another line of consequence that runs through the fabric of our relationships with other species, the line that emphasizes that so much 'tradition' is male tradition, apparent in the way many of these enterprises are conducted, defended and portrayed. Women take part but often as a minority presence. This may be more connected with lack of opportunity rather than lack of inclination but the overwhelming fact and perception is that it's boys doing stuff, being boys. It is what whalers do, what the 'hunters' of Malta and Italy and Cyprus do, as they dress up in camouflage gear to kill small birds, what 'The Men of Ness' do. The 'guga hunt' may once have been, as one defender of the practice suggests, a welcome escape for the lads of Ness from the matriarchy of their society, a result of the drastic loss of men in two world wars but that seems, at best, a time-limited excuse.

I look at photos of these activities from all over the world and they are all so starkly monochrome and male. In idle moments of wild fantasy, I imagine only women carrying them out but it's difficult. The basic premise is so fanciful and errant – not because women aren't capable of killing and pickling a couple of thousand young seabirds or wrestling with huge cetaceans, not because they might not want to but because it would be so unlikely, in the societies in which we live, that the appropriate number of women of suitable age and physical disposition to undertake any of these expeditions, women employed in ordinary jobs, taking the usual majority responsibilities for home and family would be able to arrange to have the time free from all other concerns, to spend a fortnight or a week or a few days together doing anything at all on the high seas, or on a rock in the middle of the Atlantic. Voices, as from the pulpit rise in my ears in stern rebuke: *immodest, ungodly, unfeminine*, while others whisper comfort into the ears of men, *glorious, sanctified, righteous, tradition*.

On a summer afternoon, a tweet appears on my time-line: 'Yet another vicious excuse for "tradition" destroying life,' it reads. Attached is a link to matters concerning a dog-eating festival in southern China. The veneers are thin – while claiming to be 'traditional' this festival appears to have been operating for only a very short time as an initiative of the restaurant trade. At the festival 10,000 or so dogs and cats will be killed and eaten over the period of ten days and since this particular event is a recent addition to the undistinguished history of worldwide animal cruelty, the ways in which its organizers attempt to validate it are tediously familiar.

Not long after, I read a headline that suggests half the world's donkeys may be slaughtered within the next few years so that their hides may be used for the traditional Chinese medicine *ejiao*. The accompanying text is as shocking as recalling that the practices of Chinese traditional medicine have brought about the near extinction of rhinos, tigers, pangolins, sun bears, saiga antelope, musk deer, sea horses and snow leopards, as shocking as the way many medicines are obtained, such as the keeping of Asiatic black bears for the harvest of their bile. So many donkeys are being stolen, traded and slaughtered that their numbers in Brazil, Botswana, Kenya and Ghana and other countries are being drastically reduced and communities who depend on them are no longer able to afford them. I remember those augural words: 'the decimation of the only assemblage of life we know of in the universe'.

While the demand for 'traditional' medicine is fuelling the vast international illegal trade in wildlife, questions can be asked about the roots of these 'traditions'. Although it's claimed as one, there's doubt about whether or not eating dogs really is a tradition in China. Some histories suggest that dog eating was popular until the end of the Han dynasty around 220 CE but declined after that, the spread of Buddhism and Islam making the practice less common. For many in China, particularly nomadic peoples, dogs were considered to be valuable working animals. In an early chapter of the fourteenth-century Ming dynasty novel *Shuihu Zhuan* ('The Water Margin'), the wonderfully rumbustious character Lu Zhishen, having murdered a lecherous butcher, seeks refuge in a Buddhist monastery where he very soon disgraces himself by getting drunk, vomiting over the other monks and allowing his prohibited, hidden treat of a roast leg of dog to fall out of the sleeve of his monk's robe, suggesting that for some at least, dog eating was neither condoned nor a tradition, although dogs have always been eaten and included in the armamentarium of Chinese medical resources.

The great Ming dynasty scholar of medicine, pharmacology and the natural world, Li Shizhen, widely regarded as the 'father' of traditional Chinese medicine, advocated the use of large numbers of plants and animal parts in his extensive *Compendium of Materia Medica*, the *Bencao Gangmu*, but the modern popularity of these practices may owe more to Mao Tse-tung, who enthusiastically promoted the idea of 'tradition' in medicine. This was at least in part because he regarded it as expedient and cheap for use in his campaign to encourage the 'barefoot doctors' who would provide much needed healthcare for rural China. It's notable that he relied on Western medicine for his own health requirements and that his knowledge of the natural world was exemplified by his 'Down With All the Pests' campaigns of the 1950s. By encouraging the wholesale destruction of sparrows, which allowed the flourishing of locusts, he brought about a famine which killed millions.

Leviticus, it turns out, may have had a point. Chapter 11 states: 'They shall not be eaten, the stork, the hoopoe, the bat . . . and anyone who carries their dead bodies shall immerse their garments and shall be unclean until evening.' Bats, of 1,300 species in the order Chiroptera, diverse, invaluable pollinators, pest controllers, spreaders of seeds with their unique immune systems evolved to copy with the high metabolic demands of flight and consequent ability, when placed under stress by environmental or anthropogenic factors, to initiate 'viral spillover' – the shedding of pathogens to which they are host – may have been one of the sources of severe acute respiratory syndrome coronavirus 2 (SARS-CoV-2). But it was neither bats nor any of the other creatures implicated, such as the grotesquely mistreated pangolin, who caused it. We were the only species responsible, by our wilful mishandling, trading, slaughtering and consumption of other species. The connections between the wildlife trade and subsequent spread of viruses has been known about for decades. Contemplating the fact that these practices should have been vigorously discouraged and prevented is small comfort in a world devastated.

The thought of the donkeys lingers. In every way, tradition has dealt harshly with these creatures. They've been used as beasts of burden, ready victims of our will to exploit, their name cruelly embedded in our language as an exemplar of stupidity and stubborn recalcitrance. Donkeys seem always to be there in the backgrounds of our lives, ambling in the alleyways of our consciences, ready targets of grumbling complaint in so many travellers' tales.

One such account is Robert Louis Stevenson's *Travels with a Donkey in the Cévennes*, in which he describes his relationship with Modestine, the small donkey he bought to accompany him. In his treatment of Modestine, he is harsh, petty and self-serving, referring constantly to his 'cruel chastisements' of her, to 'goading and kicking' her and hitting her with a stick embedded with a nail which makes her bleed. Although he expresses remorse, he doesn't stop. He sells the donkey after the journey is over, realizing only later the extent of his regret. 'I had lost Modestine,' he writes, '. . . she was patient, elegant in form, the colour of an ideal mouse. Her faults were those of her race and sex. Her virtues were her own. Farewell and if for ever . . .' and only then, he weeps.

8

The Hunt

We met early for a walk on a brilliant, freezing morning. We'd been looking forward to it, all five of us feeling this time between Christmas and New Year to be special, as if something nameless had been accomplished, that it was a small burst of freedom between one rigorous set of obligations and another. I drove out from Aberdeen in the quiet stillness of early Sunday, past the big houses set back from the road, the mock Tudor villas and granite mansions, past their frosted lawns and tall fringes of dark, leafless trees, beyond the outskirts into the hills of Deeside. When I stepped out of the car at the clearing where the others were waiting, it was into the scents of winter morning, of wood smoke, peat and ice. The walk we'd planned couldn't be long – the days are too short, the light too brief in late December, the winter solstice only a week past.

It was still early when the convoy of Land Rovers passed us, clearly on its way to shooting grounds further up the hill. Some of the passengers acknowledged us, waved or nodded and we nodded back in reluctant greeting towards this jolly cargo of men and guns, and walked on. It is not an uncommon sight here but for all that the day was changed, not because we weren't all well aware that 'game' shooting takes place but because we were. We carried on talking and laughing but as we did, there hung over us silently the meaningless deaths of other creatures. We stopped for a chilly picnic by a high cairn looking down over the calm landscape of shaded green, of light on water, birch, larch and pine. By now, we all felt ready to face the long haul of a Scottish New Year. Nothing about our day was destroyed by our seeing those men except the frail hope we sometimes have that people might find other ways to spend their time.

Driving back through the small roads of Deeside in early afternoon, I couldn't help wondering about the day those men would have had, their incomprehensible pleasure in killing and although I hadn't thought about it for a long time, I remembered the evening I spent many years ago at the New York headquarters of a British gun manufacturer. The gun company wasn't somewhere I'd chosen to go – I was working, helping a family member, part of whose job it was to promote British industry in the United States. The evening was one in a particularly rainy November. I recall the sensations: the heavy rain, negotiating the late-afternoon crowds of Manhattan while gripping tightly to the huge umbrella the hotel had lent us. We'd been working in glamorous places, art galleries and expensive hotels so that there was a distant, filmic aspect to the days that made them feel as if we were acting for an invisible camera. Everything was far removed in every way from my usual daily life of child- and pet-rearing, of housekeeping and writing, everything quotidian but vital. For the time that I was there, I was a willing helper, anxious to do all the menial but necessary things, the dogsbody tasks that secretly I enjoyed; preparing the gifts we'd give to the attendees at the end of the evening, collecting invitations, carrying armfuls of lightly scented mink coats to be hung carefully on hangers. (Fur coats seemed to a European worryingly anachronistic.) I love New York and working rather than just visiting, even for a few, too-short days added a dimension – if not exactly of importance – then purpose to being there, giving me a sense of having some small degree of meaning in the place.

Our destination, the headquarters of the gun manufacturer, was only a few blocks away from our Manhattan hotel, set among the most exclusive shops, offices and banks. Their building was imposing but anonymous. If you hadn't known as you walked past it, you might have wondered what went on there for nowhere did it indicate what was being offered or sold. You'd have assumed that it was a discreet bank for the über-wealthy or the headquarters of the kind of secretive mega-institution that without our knowing controls governments and rules the world. I didn't know anything about the company except their name, which I recognized from somewhere – from a reference in literature, Isabel Colegate perhaps, or one of the Mitfords, I can't remember now. I knew that they made guns for 'game' shooting, and that we had to be there that evening and what had to be done, but more than that, I didn't know anything at all. Inside, behind the discreet, expensive exterior, was a discreet, expensive interior. Nothing disclosed what they did or sold. They could have been selling anything – vastly expensive jewellery, art works, miniatures, icons, coins, things which would silently be carried from strongrooms or vaults to be displayed along the antique desks and tables where we were laying out our press packs and brochures. Being New York, the guns and ammunition were stored out of the sight of passers-by on the not-so-mean streets of Midtown Manhattan, kept prudently on the fourth floor, lined up in wood and glass cases or in safes in rooms with heavy locked doors marked 'gun room'.

There's nothing in my own life more alien than hunting. With the introduction of dietary laws, hunting couldn't be carried out in the Jewish world of the early Middle East and, broadly speaking, never has been by Jews since. In a remarkable strand of cultural continuity it's still regarded, tacitly at least, as being something we just don't do, not expressly forbidden but bearing a legacy of disgust and contempt for the murderous behaviour of Nimrod and Esau, as well as awareness of a number of laws regarding the correct treatment of animals. It's unsuitable for *rachamim b'nei rachamim*, 'compassionate children of compassionate parents' (which I wish we all were), seen as frivolous, cruel, damaging to the character of those who carry it out and an essential bar to entering 'the world to come' (although on the whole, we're pretty vague about what's on offer in 'the world to come' and distinctly short on mortality bromides).

Until that evening, I don't think I'd had any feelings about the manufacturers of 'sporting' guns, if I'd thought about them at all. I had thought a lot about warfare and arms, the meaning of conflict, the effects on human life because who can do anything else? When I was born, the shadows of the last world war hadn't yet dissipated, which I know now, they never will. As a teenager during the Cuban Missile Crisis, I demonstrated in the streets of Glasgow against nuclear weapons, attended talks, read books – John Hersey's *Hiroshima*, Robert Jungk's horrifying *Brighter than a Thousand Suns*. For as long as I'd thought about it, most of my feelings about guns, killing and warfare had been directed towards the 'military-industrial complex', the armaments manufacturers, the ones who make the big stuff, Bob Dylan's 'Masters of War', the creators and facilitators of international warfare, the dealers in cluster bombs, phosphorous bombs, nuclear weapons and, ultimately, the splitters of atoms. Robert Oppenheimer's weirdly detached description of the satisfaction to be found in the work of having created the bomb – 'technically sweet' – lurked somewhere in my mind, this dissonant oxymoron, centring my thoughts on the contradictory and the absurd, everything characterized by the para-doxical concepts surrounding what we as humans consider acceptable to do to one another.

Later, as a student living for a few years in Israel, I found myself among many sorts of war. During those years I travelled two or three times to Arad and Dimona in the northern Negev, the places where even then everyone knew that nuclear weapons were being made and stored. My only experience of the actuality of guns, apart from seeing them carried everywhere by the army was when, during university holidays, I was obliged to do kibbutz guard duty and in the course of long and boring nights my fellow-guard taught me to take apart and reassemble an Uzi. It was easy once I'd been shown but I was all too aware of what these assorted, heavy pieces of metal could do. I learned what it was to be on the other end of a gun when, as a resident of occupied Palestinian Jerusalem, I became subject to random checks and curfews, then a few years later in Lebanon during the civil war, when I had to scuttle quickly inside away from the dry, sharp sound of sniper fire.

Before the guests arrived, one of the firm's employees, a young American, offered to show me around. He led me upstairs to an elegant room more reminiscent of a library than a gun showroom where glass cabinets enclosed rows of guns of polished wood and finely engraved metal. The room felt like a bit of Britain transported to New York – the photographs on the walls mostly of Scotland and the more or less familiar figures of people who must have been the firm's customers, men in tweeds carrying guns purposefully across unnamed heather moors. When I looked closely, I could see from features of the land-scape that they weren't far from where I live, the known contours, the wide north-eastern skies. An enthusiastic employee, eager to show off the beauty and efficiency of his products, tried to explain to me what it all meant, 12 bore or 20 bore, about the weight of guns, what double triggers did, spring blades and stocks and bolt action. He talked of the walnut and metals used in the manufacture, the painstaking adorn-ment, the ancient techniques of Damascene metal. You could if you wished, he said, have the image of your favourite gun dog engraved on the lock-plates of your gun.

'A lovely memorial,' he said, 'after they've passed.'

I looked down from the window at the rain. The tides of water rippling down the long windows of that beautiful Midtown building had muted the almost-dusk light, turning the view of the streets below into an Impressionist landscape, a blur of faint traffic moving against the undulations of dark umbrellas. I wondered what I was doing there, among these objects designed for the members of one species to kill those of another.

The sensations of the evening were contradictory, the ease and comfort of the lives the place represented set against the actions and desires of their purpose. We were there, surrounded by refined luxury and high cost, bathed somehow in the patina given off by those things which are perfect but unnecessary, the ones that might make people ponder on the quality of everything else in the lives of those who are able to spend large sums on things to wear, give as gifts, celebrate or kill.

At some time during the evening, I found that there was another rebel present, a fellow-subversive, a freelance photographer there to take photos of people drinking champagne and Scotch. We discovered our shared dissent in the way you do, from exchanged glances and muttering in corners, the natural habitat of the disgruntled. It was a relief when finally we had a chance to make clear to one another what we thought, employing effort to look sufficiently casual, as if we weren't using the words we were. Our conversation itself seemed like an attempt at self-exoneration. Only the line of duty had placed us there. Of course, we said to one another, we wouldn't have dreamt of being there otherwise although we didn't really need the mutual reassurance. I hadn't asked the young employee about the cost of guns. It would have seemed impolite and possibly vulgar and anyway I didn't need to. No one without the spare sum of money which, if it hadn't been spent on buying a gun, could have paid for a house in most places in the world, would have been there out of choice, ever. Nor anyone who had ever thoughtfully considered the concept of 'game' in conjunction with the sole purpose of causing death for sport. Everyone talked and drank and perhaps did business. Perhaps, as they were meant to do, they exchanged promises, prospects, invitations. At the end of the evening, we collected ourselves and all our bags and boxes and left and the heavy doors closed behind us as we went out into the New York night. In a few days, I was back in Scotland, in Aberdeen, a few miles from the moors where those guns are used. Although I didn't have particular cause to think about them or their purposes, I did at odd times in the lees of that evening, which stayed with me. A taste of the sinister lingered with the memory, connecting me with something ineffable, the scent of power perhaps, or a bitter knowledge of the things people do for sport.

Now, as I drove back into the city on that late December afternoon, lights were being switched on in houses along the road. The sky was streaked with the purple light of approaching snow. I thought of everything I'd discovered or learned or been taught in the years since that evening in New York. It felt like a moment enclosed within a long loop of time and as I think of the calm light of that douce afternoon it seems like another strand, more rhythmic clicking of the abacus, one more line of consequence.

Hunting's always been with us, evolving over millennia from necessity to recreation, from the provision of food to the sport of gods, of Artemis and Odin, Actaeon and Woden – to the sport of kings. In time, hunting became an activity often carried out for the enjoyment of the rich and aristocratic in their exercises of power over others, human and not, and in the affirmation and extension of their possession of land.

In the twelfth century, the part of the city where I live was the king's hunting forest, alive with fox and wildcat, lynx and boar. Properly urban now, the only traces left are in the whispers of its names, in kings and queens, woods, forests, dens. In her book *The Writing Life*, Annie Dillard points to the long association of hunting with the aristocracy, describing one hunter's daily schedule as being particularly appealing, that of the late-nineteenth-century Danish aristocrat whose day on his estate was spent hunting grouse and woodcock, swimming, chatting with fellow hunters before bathing and changing into white tie and tails to enjoy an evening of dining and smoking cigars. The aristocrat was Wilhelm Dinesen, father to Karen Blixen.

The word 'hunting' itself seems remote now, when so little of it actually involves hunting. The vast majority of modern hunting isn't done for food or necessity of other sorts such as the management of species populations, and where it is still done for subsistence it's not accompanied by the large-scale spending that takes place elsewhere, by vehicles, equipment, flights, the services of staff, keepers, beaters, rangers required to facilitate the pleasure some gain from the slaughtering of other species, much of it specially bred and raised for the sole purpose of being shot.

In a paper about hunting, the author asks whether or not hunting is 'part of being human'. The question seems odd, more uneasy self-interrogation than question, an awkward meeting-point of the rhetorical, the angsty and the self-justificatory. The only answer required seems to be that if hunting's a necessary part of being human, then it's just part of being us, part, possibly, of the long evolutionary pathway of our existence and therefore beyond the criticism of anyone who might think it inhuman or even inhumane. Wondering if something or other makes one human seems little more than an expression of ontological insecurity about what one's doing in the world, a concern about one's possible tendencies and what – heaven forbid – those might allow, or even oblige one to do. (Which bit of being human? The eating part? The killing part?) It's not difficult to explain why justifications might be necessary. Often, rightly or wrongly, hunting is seen to represent things with which many might not want to be associated – extremes of privilege, a dangerous admixture of wealth and power, avoidable cruelty, unabating misogyny, purposeless violence, a contempt for threats to the environment and planet, yet another obsolete defence of 'tradition' in the driving of many other creatures to near extinction for our own recreation. By now, too much on which hunting is based, too much that surrounds and nurtures it – from the jarringly expensive, unnervingly sophisticated weapons and equipment it uses – seems out of place and out of time.

In his brilliant book *A View to a Death in the Morning*, the anthropologist Matt Cartmill wrote of the ideas that the role of hunting is based on, including the 'hunting hypothesis', an idea traceable to the Australian palaeontologist Raymond Dart who, while working in South Africa in the 1920s, interpreted bone evidence from the Transvaal as proving that the species he described and named – *Australopithecus africanus* – used bones as weapons to carry out horrific violence and murder. In a paper in 1953, Dart described them: 'slaking their ravenous thirst with the hot blood of victims ...', 'greedily devouring lurid writhing flesh' and elsewhere writing of the 'blood spattered, slaughter-gutted archive of human history'. Dart's description created a direct connection between his interpretation of early violence with contemporary human behaviour, what became known as the 'killer ape theory'. Although much criticized by fellow-scientists both then and now, his view proved popular. Perhaps because of his wild and overdramatic presentation, Dart's ideas were taken up by the playwright Robert Ardrey who designated them 'the hunting hypothesis', writing in his book *African Genesis* in 1961 of a human 'genetic cultural affinity' for weapons. Both Dart and Ardrey suggested that, as potentially violent murderers, we are all 'children of Cain'.

Despite later demonstrations that Dart's analysis was wrong, that he had misinterpreted the evidence, that early humans were more prey than predator and any meat they ate was probably scavenged, the ideas prevailed. During the 1960s, books on human origins and behaviour proliferated, many of them presenting explanations derived from the distant past, possibly, as many have suggested, in an attempt to understand how the worst acts of the Second World War could have been committed. One of the most popular of these was by the Nobel Prize-winning ethologist Konrad Lorenz, a man greatly admired and emulated by Ardrey. In the bestselling *On Aggression* Lorenz wrote: 'there is, in the modern community, no legitimate outlet for aggressive behaviour. To keep the peace is the first of civic duties and the hostile neighbouring tribe, once the target at which to discharge phylogenetically programmed aggression, has now withdrawn to an ideal distance, hidden behind a curtain, possibly of iron.' At the time and since, few people seem to have questioned his suitability as a social commentator. I love his book *King Solomon's Ring* but although Lorenz may have been engaging when writing about jackdaws and his early life in Altenberg, he was less so in his application to join the National Socialist German Worker's Party (NSDAP) when he wrote that his 'entire scientific life's work stands in the service of National Socialist thinking' or that the 'purity' of the human species was being threatened by 'invirent types' – the domesticated and urbanized – who would, if allowed, 'penetrate the Volk body like the cells of a malignant tumour' – those who would, then and now, be called 'rootless cosmopolitans', 'citizens of nowhere'. It wasn't mere rhetoric – during work as a military psychologist, Lorenz took part in 'racial harmonization' selections in Poland in the early 1940s and continued to have an unaltering and dangerous admiration for the idea of the 'pure'.

Demonstrating similarly distasteful views, Raymond Dart is reported to have walked out in fury when, at a conference in South Africa in 1929, a time when the foundations of apartheid were being laid, the British archaeologist Gertrude Caton Thompson demonstrated convincingly that the Iron Age city of Great Zimbabwe, which she had been excavating, was built by indigenous people and not, as had long been believed, by the Queen of Sheba.

Ardrey, reporting for an American magazine on the bitter, anti-colonial Mau Mau uprising in Kenya in the 1950s, wrote of: 'the primal dreads of a primal continent . . . Africa scared me. If this continent had indeed been the cradle of humankind, and had I been the first man, then I should have been born in fear', expressing his profoundly racist view while overlooking the fearsome and brutal iniquities of British colonial rule and its violent suppression of the uprising.

What seems remarkable now is that the 'hunting hypothesis' and its long and lasting offshoots could be used as justification for anything at all. Anyone might, after all, want to pick lice from a stranger's hair, walk naked in the streets or pick up a large stone and bash their neighbour over the head but most of us don't, despite the unprovable belief that one or other ancestor may have done it before us and the reason we don't is because other elements have intervened between us, the ancestor and the urge: diachronicity and the processes of socialization, taboos, morality, the development of our brains, the belief that we have, in some respect or another, progressed, changed and that doing what we did once because we once did it is not a good reason for continuing to do what's damaging, antisocial, cruel or might be regarded as simply wrong.

'What "play",' the evolutionary biologist Stephen Jay Gould wrote, 'would evolution have if each structure were built for a restrictive purpose and could be used for nothing else? How could humans learn to write if our brain had not evolved for hunting, social cohesion or whatever, and could not transcend the adaptive boundaries of its original purpose?'

Still often used in justification and explanation, *fons et origo*, 'the hunting hypothesis' provides a soothing refuge for those who kill for pleasure or those who, as well as wondering if hunting makes them human, often express their devotion to the practice by writing of it in semi-mystical or religious terms as a way of reflecting their 'evolutionary roots', a way to 'transcend the bounds of human consciousness', as 'a mystical union', as 'numinous', 'a sacrament', 'a Dionysian moment', as a 'holy occupation'. Many like to believe that: 'Hunting is stewardship of the land', 'Hunting is tradition', 'Hunting is the way we live sustainably with nature' or 'It's natural for Man to be top predator' (just to prove to social Darwinists everywhere that they're not forgotten). The word 'Pleistocene' occurs a lot. On a website devoted to studies in ecology and nature, I find a splendidly opaque statement on hunting written by a philosopher, suggesting that without humans, hunting wouldn't exist because no one would recognize it as hunting.

The principal claim of the 'hunting hypothesis', that the urge to hunt is innate in humans, fails to address the question of why if this is so such a small and diminishing percentage of people actually do it. The American historian and hunter Jan Dizard questions the idea of hunting being an essential feature of humankind, writing that if he didn't hunt he wouldn't be going against his nature or 'defying' his genes.

It was when I was with a visiting friend in a tourist shop in a town to the north that I came across a greetings card which stood out from others on a stand of cards of misty Highland scenes and flower prints. On the front, a stark black-and-white photo of a man with a rifle, on the back the quotation: 'One does not hunt in order to kill, on the contrary, one kills in order to have hunted.' It was it said, from the book *Meditations on Hunting* by José Ortega y Gasset.

After I read the quotation, I wondered what it meant. Why should anyone reduce the act of killing to a role secondary to their own pleasure in doing it? Could he really be suggesting that the purpose, or pleasure for the hunter, is supreme, the killing of another creature merely a lesser token point in some onanistic, anthropomorphic act of self-fulfilment? Ortega was a distinguished Spanish philosopher who was living in Portugal during the war when he wrote this book. Still regarded as a 'classic' of hunting literature, it is described by hunting organizations, magazines, individuals and commentators in near reverential terms as a work of insight and vision, lauded in a series of introductions and prefaces in the edition I eventually bought, a reprint by a Montana publishing house devoted to books on outdoor pursuits, with the usual motif of hunting on the front cover – a vaguely phallic depiction of a man with a gun.

'Hunting is what an animal does to take possession, dead or alive, of some other being that belongs to a species basically inferior to its own,' Ortega wrote in 1942. 'In our time – which is a stupid time – hunting is not considered a serious matter ... The most appreciated enjoyable occupation for the normal man has always been hunting' and 'In this way hunting resembles the monastic rule and the military order.' Ortega's view was uncompromisingly patrician and xenophobic. He despised the utilitarianism of the hunter-for-food and regarded hunting as the proper exercise and reward of the aristocratic and privileged, sharing with Lorenz the view that domesticated animals and urban humans are similarly degenerate and impure, and with Dart and Ardrey the tendency to blame it all on Cain. The foreword to my edition of the book, a down-home 'practical wisdom' accolade written by the American diplomat and fisherman Datus Proper includes a brief anti-communist manifesto which declares that one of the benefits of hunting is that it leaves its more proletarian participants no time for doing disruptive things like starting revolutions.

I don't often annotate books. I did once or twice as a teenager when I pencilled a few penetrating aperçus into the pages of my copy of *Palgrave's Golden Treasury* but the need didn't strike me again for a long time. There's something about doing it that erodes respect – like any noxious habit, start and you may not stop. Anne Fadiman in *Ex Libris* discusses the different ways in which prolific readers treat books, either ultra-respectfully, being careful to leave no trace on pristine covers and pages, what she describes as the behaviour of 'courtly lovers', or 'carnal lovers' who read books in the sauna, shred and damage pages in the act of reading, or, as she describes one friend doing, 'suck them like a giant mongoose'. When I annotated *Meditations on Hunting*, it was more feral than carnal, stars of the lurid pink pen, which I used on the pages, illuminating the incidence of absurdities and offences, the few references to women, all demeaning.

It's easy to dismiss the book as out of date and irrelevant, which it is, but the place Ortega describes lives on unchanging in the predominantly male hunter's world. There are huge differences between the way American and European hunting's carried out, but the basic elements are present in both: the centrality of guns and the semi-military, the predominance of men and similarities in the ideas and ideologies which support and encourage it.

One hunting magazine offers me the tempting proposition of buying a car window decal declaring 'Testosterone, GET IT ON' over a picture of a large, antlered creature. 'There isn't too many women I know of who practise hunting, it's a thing of men,' someone writes in another hunting journal. The idea of hunting as an integral expression of masculinity is central to a view in which the belief in testosterone as the major determinant of sex difference, the most important driver of male behaviour, lurks behind every validation of the practice. The word 'chromosomes' occurs frequently too in a routine synecdoche of self-justification: 'Only when our culture accepts the needs of living men as shaped by a prehistory which is still urgent within them, communicated to each by his chromosomes, will we be ready to follow the lead of Ortega's *Meditations on Hunting*,' the 'deep ecologist' Paul Shepard wrote in his introduction to the book. In *Testosterone Rex*, Cordelia Fine suggests that far from fulfilling that particular role, testosterone is only a part, a far more minor one than commonly believed, of the complex and interrelated structures of human biology and behaviour.

In the library, I pick up a new book about Ernest Hemingway and open it. There, predictably enough, is a photo of him after a duck shoot, after trying out his new shotgun, eighteen ducks limp in the boat. There's always a photo, a gun, something dead, a pile of antlers. In a splendidly witty essay about books best avoided by women, Rebecca Solnit writes: 'Ernest Hemingway is also in my no-read zone, because if you get your ideas from Gertrude Stein you shouldn't be a homophobic anti-Semitic misogynist, and because shooting large animals should never be equated with masculinity. The gun-penis-death thing is so sad as well as ugly . . .'

Indeed it is, but nobody seems to have told the many men who write books and articles about hunting who perhaps do not realize that to write of hunting as 'romance', 'seduction', 'ecstatic consummation', 'sexual release', 'possession', or of the animals they're hunting as 'love objects' or 'desirable' tells the world something fundamental about them. In hunting literature, references to the sexual abound: 'Hunting includes killing like sex includes orgasm. Killing is the orgasm of hunting,' the hunter Ted Kerasote writes. Paul Shepard saw sex and hunting as manifestations of 'venereal aggression' and believed that 'the association of menstrual blood and the idea of the bleeding wound is inescapable'. Ortega too seems to have had difficulty with the idea of menstruation: 'A white rag stained with blood is not only repugnant, it seems violated, its humble texture material dishonoured' although he did get all stirred up by the thought of a pack of hounds in chase: 'Suddenly the orgiastic element shoots forth . . . which flows and boils in the depths of all hunting . . . Things that before were inert and flaccid have suddenly grown nerves and they gesticulate, announce, foretell.'

In a deeply unpleasant and disturbing account in his book *Bloodties*, Kerasote described an erotic dream encounter in which an elk presents herself to him in the guise of a naked woman, 'her eyes wet and shining'. A British book on hunting, which someone gave me as a present, expressed similar ideas in more British terms, talking about what 'chaps' do when they're feeling 'clubbable', musing on the aphrodisiac effects of the power and money needed to be able to take part in 'game' shooting, except that money's referred to coyly as 'wherewithal'. 'Sex and death, that's what it's about. After all, who can resist a successful hunter-gather?' the author asked. Shooting specially bred birds with expensive weaponry is not 'hunter-gathering' – it's the pursuit by those who value themselves too highly of those they value too little.

There's a creepy phallocentric, homoerotic concentration too on size, on antlers, bucks and horns, on 'boosting your bag', on numbers of 'kill', on 'penetration', a word some hunters with crossbows seem to like – 'Bow hunters get better penetration,' one company selling a selection of sophisticated crossbows boasts. The results of this preoccupation find horrible expression in 'trophy' hunting, in the artificial, overbreeding of deer with preternaturally large, hideously disfigured antlers to provide bigger 'trophies', in the killing of the largest possible lions or buffalo and in the indiscriminate destruction of raptor species which prey on 'game birds', threatening to reduce 'the bag', a practice which has drastically reduced numbers of birds of prey in the UK.

In *A View to a Death in the Morning*, Matt Cartmill suggests that many hunters explain the activity of hunting much as rapists do, by claiming that their victim was 'asking for it', illustrating this with Ortega's description of hunters coming upon some wolves:

> It is not man who gives to those wolves the role of possible prey. It is the animal which demands to be considered in this way ... Thus they automatically convert any normal man who comes upon them into a hunter. The only adequate response to a being that lives obsessed with avoiding capture is to try to capture it.

This is not to suggest that hunters are rapists but simply to say that the language and assumptions of much hunting literature ring eerily familiar in the wearingly unchanging male narratives of entitlement, subjugation, violence and death. In his article on bird hunting in the Mediterranean, Jonathan Franzen quoted an Italian former hunter turned WWF founder as saying that the Italian devotion to hunting is part of their 'attitude of virility', describing the practice of using inappropriately over-powerful ammunition to kill small birds as being 'like a rapist who loves women but expresses it in a violent and perverse way . . .'

In her powerful re-casting of this strand of male hunting literature, Susan Griffin writes in her book *Woman and Nature* of myth and reality, describing in juxtaposition to the narratives of object and desire the starkness of the hunt:

He is burning with passion . . . He will make her his own. She reveals only part of herself to him. She is wild. She flees whenever he approaches. She is teasing him. (Finally, she is defeated and falls and he sees that half of her head has been blown off, that one leg is gone, her abdomen split from her tail to her head and her organs hang outside her body. The four men encircle the fawn and harvest her too.)

There are 'post-kill rituals' too, the 'blooding' – the smearing of animal blood on one's own or another's face, and the ones described in one hunting magazine – the 'chest thumping', the 'laughing and hollering', the eating of the raw liver of the animal, the slitting of its throat.

The 'ethics' of the practice are frequently discussed in hunting literature. In Ortega's chapter on the subject he wrote: 'At times, killing the enemy, the madman, the criminal, is obligatory and unavoidable.' Elsewhere, the ethics are confused with the etiquette and norms of accepted behaviour, often including the concept of 'fair chase'. Described by the American hunting organization the Boone and Crockett Club (yes, *that* Boone, *that* Crockett), which was formed in 1887 by Theodore Roosevelt, who while contributing hugely to conservation and land management in the United States, was also a considerable slaughterer of wildlife, 'fair chase' in hunting must be: 'ethical and sportsmanlike' and carried out: 'in a manner that does not give the hunter an improper advantage over such animals'.

Having a gun tends to give the hunter an advantage. British guns, limited by strict control laws, are shotguns and rifles, 'side by side', 'over and under', 'hammer ejector', 'bolt action', the more expensive ones finely made – like those of the gun company I visited – elaborately engraved with stocks of Turkish walnut. American guns, their manufacture relatively unfettered by the law that determines their use and availability, offer technical features which dazzle – digital displays, sensors, onboard shot computers, trigger-guiding mechanisms, microprocessors, electro-optics, Wi-Fi and more. I read that one particularly sophisticated gun is ideal for: 'hunters who don't have the time to achieve the highest levels of skill who can now make dreams of a trophy kill a reality'. You can, in other words, kill an animal from so far away that you can hardly even see it. Another is described as 'severely precise', but far from their image as clever, brilliantly engineered marvels, these weapons, paradoxically enough, render their users skill-lessly impotent, mere adjuncts to a piece of equipment, the tool of a tool, the truly perfect 'bête machine'.

The long association between war and hunting extends beyond guns and violence, not least in the laying down of regulations, of 'fair chase' and 'the laws of war', both expressions of the same airy hope that some of the humans involved in the enterprises will behave better than they do. One discourse on 'fair chase' includes a list of how a hunter should behave: Don't put out feeders then sit in tree houses or on 'high seats', waiting. Don't kill an animal in water, in deep snow, on ice. Don't kill animals inside fenced enclosures. Don't use electronic devices, spotlights, poisons, tranquillizers. Don't use planes or drones. Don't shoot from vehicles. Don't drink or take drugs, don't shoot at things for the hell of it, don't make a mess, leave stuff behind. Obey the law. Know what you're doing. But many don't know what they're doing, the ones too busy, perhaps, 'to achieve the highest levels of skill'. They wound birds and animals, leave them for dead, often 'maimed, mutilated, despoiled' as Joy Williams describes in a fiercely critical article, 'The

Killing Game'. The advice in more than one British hunting book I read is to collect up the wounded birds where you can and 'dispatch them discreetly'. Of the hundreds of thousands of birds shot in Britain during the shooting season it's estimated that 40 per cent aren't killed outright.

There are many reasons why people hunt. Some engage in quiet, small-scale hunting, the kind a friend explains to me. He is someone more knowledgeable about birds and conservation than anyone else I know and occasionally shoots a duck or some other bird to cook as part, he says, of walking in the hills with a friend or of spending the day alone by a lakeside. There are those who hunt to avoid the products of the 'factory farm' but choose to eat meat. There are those who kill what they need and no more, who do not do it for pleasure alone. One of my favourite characters in Alice Munro's work appears in the short story 'A Real Life', the solitary, modest hunter Dorrie Beck, a small-time trapper of muskrats, groundhogs and rabbits, a woman after whom a horse was named, who loves: 'the spring watercourses, the system of creeks she followed, tramping for miles day after day, after the snow was mostly melted but before the leaves came out, when the muskrat's fur was prime.'

Women hunt too, although fewer than men. They too pose with their bloody trophies, dress up in camouflage, smear their faces with an animal's blood. Some say they've taken up hunting so that they can eat hormone-free, non-industrially farmed meat from animals they've killed themselves. Still others say they do it as an act of rejection of the male dominance in hunting. Some American hunting magazines for women are hardly less sanguinary or stereotypical but they include features on how to look after your jewellery when hunting ('marriage can be difficult and you'd rather not lose this very special symbol . . .'), how to use burnt wine-bottle corks as camouflage face-paint, how to spring-clean taxidermy and how to cook 'venison mac'n cheese'.

In the anthology *Women on Hunting*, writers, both participants in and critics of hunting, thoughtfully examine their own experiences. For the editor Pam Houston, hunting has given her a 'deeper understanding of my animal self. I also have the blood of five fine and wild animals on my hands and will never forget it.' In her poem 'She Who Hunts', Nance Van Winckel leads us through the quiet processes of the lone hunt and the aftermath of memory, wondering how long it will take for her to forget the act of killing.

Fathers, husbands or boyfriends are often cited as the primary influence in a woman taking up hunting, often in order to ensure the continuation of family tradition or as a form of homage to a dead father. Terry Tempest Williams, in 'Deerskin', considers her father and brother's deep knowledge of hunting, their learning from the oral traditions of the Navajo, 'a model for ecological thought expressed through mythological language' while Tess Gallagher, feeling herself part of the complex hunting traditions of both men and women in the Northwest, decries the 'male anger and wrongheadedness and escapism' now involved in an activity she sees as having lost the sense of seriousness and purpose it once had. Elsewhere, the writer on religion and philosophy Jill Carroll explains her feelings about being a hunter, of hunting to eat and of killing the animals she eats herself as being the most ethical way of being a carnivore: 'When I take to the field with my shotgun to hunt, I say "Yes" to the world as it is, and to my place within it,' she says. There are echoes of Mary Oliver's 'miraculous interchange', as she describes the continuity of consumption and renewal, in a context which defines hunting as a part of the entirety of an organic world, a different world, and view, from other kinds of hunting which are defined only by the meaninglessness of their violence.

A few years ago, a man shot a lion in a national park in Zimbabwe. The creature, a twelve-year-old male, died rather slowly and his death became a cause célèbre because he was popular with tourists, named, satellite-tagged, and had for eight years been the subject of study for a conservation research team at Oxford University. The man who shot him was a dentist from a place in the United States whose name was appended to every news report as if to underline the very un-remarkableness of the place and of a man who had done nothing remarkable at all except to kill a creature he was not going to eat, with a weapon that gave him the absolute advantage of being able to carry out his egregious deed without any possible danger to himself. His sole achievement was in making himself both hated on an international scale and an object of derision. The fact that he contributed inadvertently towards the subsequent raising of large amounts of money for the conservation research unit was a small pleasure for observers but one which could not alter the malevolence of his deed or the thinking which motivated it. The man has been described as 'an avid trophy hunter'.

'Trophy' hunting involves the killing of large animals often in poor countries by wealthy people from the United States, Europe and the Middle East and the retention of parts of these animals as display items, attempts perhaps to suggest heroic qualities hitherto indiscernible in their killers. Often presented as being of both conservation and economic benefit, the evidence suggests otherwise. A study in 2017 demonstrated that trophy hunting, with its emphasis on killing male animals with particularly favoured characteristics such as horns, antlers and tusks may threaten populations. Trophy-hunting organizations in the United States auction rights to hunt specific animals. The life of a critically endangered black rhino was auctioned for $350,000 in 2014. Animals are hunted with high-velocity weapons, from vehicles, from helicopters. There are no controls on the way they are treated, before or during their slaughter. Many territories where trophy hunting takes place are subject to corruption, bad governance and the resultant proliferation of poaching. Companies who operate the tours and 'safaris' are often international, part of global systems that repatriate wealth and limit the benefits to local communities. The economic benefits to the communities in which trophy hunting takes place are difficult to evaluate and much of the evidence contradictory, but a study undertaken by the International Council for Game and Wildlife Conservation and the United Nations Food and Agriculture Organisation suggests that only 3 per cent of revenues from trophy hunting end up in local communities.

There is an environmental toll too in the long-distance travel and provision of 'luxury accommodation' and 'international cuisine', the sprung mattresses and laundry services, the bottled water, fine wines and sundry accoutrements which would-be trophy hunters demand. Animals may be raised in 'breeding facilities' for the sole purpose of being 'hunted'. Far from sending a message of conservation, trophy hunting grants the freedom to kill wantonly, to turn animals into commodities, ornaments, pabulum for the self-indulgent and vain. Displaying animal parts may be integral to someone's long cultural heritage, an echo of somebody's ancient desires to display success and virility, but it's repulsive too, past its time, part of the long tradition of imperialism, colonialism and the attitudes that accompany them, none of them long enough past to be forgotten.

The dentist posted photos of himself with the dead lion as many others have done with their prey before and since. The American woman who kills goats on the small inner Hebridean Isle of Islay posts pictures of herself in full hunting outfit. Eric and Donald Trump Jr., enthusiastic 'big game hunters', were famously photographed with their kill – a leopard, an elephant, a buffalo. One of them holds up an elephant's severed tail. They smile, as they all do, bathed in the full glow of their malignant vacuity.

And now, it's August, the time when in Britain the 'season' begins, the day when particular birds may be shot again – from the 12th of the month, red grouse and snipe. The day is commonly referred to as 'the glorious 12th' although apparently, among 'shooting men', it's just called 'the Twelfth'. By now, the weather has exacted its toll.

In most of Europe and in other places in the scorching world, the summer has been long and burning. Here in the north, it has been less so and by now what warmth there has been is in abeyance. It grows cooler. The days are alternately brilliant with sunlight or dark and overcast and autumn already seems near. Spring was cold this year, an undifferentiated season that dwindled from wild, cold winter to merge unremarkably into summer in an overlap of grey and rain. August, with long days of flitting sunlight, concentrated moments of true heat, occasional rain and wind tearing through the branches and new clusters of rowan berries glowing on the tree in the garden. The 12th dawns and ends, successfully or not, depending on your expectations. Poor spring weather may lead to low grouse numbers, with fewer available for shooting because chicks can't survive the cold. In spite of that, some do. They survive and fly and are shot and at least some of those not taken away by their killers will be sent off to game dealers to be sold.

This year as every year, the London food shop through which I once sent someone cheese as a gift emails me with their offerings – first grouse and later, as the shooting season opens for them too, pheasant, duck, woodcock, partridge. The images loom inescapably from my screen, carrying more than the simple facts of their being there, plucked, eviscerated, trussed, lying neat and parallel on white waxed paper, the maroon-dark bodies of birds somebody has shot. The photos and text make it seem simple – just a few wild creatures killed for food. How, looking at these small featherless birds, could you ever imagine the processes it took to get them there. The company tells me that these 'game' birds and animals thrive in wild, empty places, grazing on natural grasses and flowers, on berries, shoots and plants. Beside each, there's the brief warning: 'May contain shot.' This year, grouse are being sold individually. The weather has caused a shortage. I look at them and know that they come from another realm, another culture, victims of an alternative view of the value of others' lives.

It's not only red grouse and snipe who can be shot legally from this day but ptarmigan too, 'snow grouse', those beautiful high mountain birds, seasonal dressers in their grey and brown of summer, invisible against heather, rock and lichen, pure white in winter against the deep, bright snow.

Once, I might not have questioned that these are 'wild birds which feed on natural diets of acorns and berries' as the company claims but now I know better. Far from the description, far from the 'natural', from the flowers and berries and wild grasses on which they're meant to feed, these small birds have been imported from factory farms abroad either as poults or eggs. Estimates of numbers vary, although there may be as many as 60 million partridge and pheasants imported into the UK transported from countries throughout Europe, reared in overcrowded conditions, often with extreme cruelty before being released in order to be shot. There are concomitant, although as yet uncertain, effects on the environment and on resident and migratory species. The birds being shot may be wild birds, as grouse are, protected from predators by assiduous gamekeepers who shoot, trap and poison populations of mammals, corvids, raptors, some legally but many not, fed by way of boxes placed around the moors with medicated grit containing anthelmintic drugs to combat the parasitic nematodes, which they may succumb to, raised on land the majority of which is expensively bought, inherited, privately owned, heavily subsidized, maintained and managed, heathered upland which has been systematically, rotationally burned to provide these birds with their optimum habitat for breeding.

This care, this expenditure, this almost wholesale destruction of other forms of wildlife and eco-systems, this damage to rivers and the life of rivers, to carbon storage, this encouragement of flooding downstream, has all been undertaken for the sole and exclusive purpose of keeping these birds alive in order to allow 'driven' grouse shooting, the sport of choice of the ultra-rich, carried out because grouse are prized, because they're wild, because, frightened suddenly from the ground by 'beaters', they rise, fly low and fast, in their panicked flight becoming no more than challenging, swift, moving targets for the line of waiting guns, an estimated 700,000 or so killed each shooting season.

It's difficult to know how many are shot because they don't have to be recorded. Huge numbers of released imported birds die through predation or starvation, being run over, through worm infestation, protozoal cysts, disease, hepatitis and enteritis encouraged by the crowded conditions in which they're kept. On some shoots, so many birds are killed that they aren't even taken away for food – one estimate is that around half are not eventually eaten. They're shot, dumped, shovelled into 'stink pits', discarded by roadsides and thrown into dustbins, victims of a system in which they're no longer themselves, no longer living creatures with origins, history, complex behaviour, habit and inclination but things only to be shot. They've been turned by shooting and the language of shooting into units, transferable to money by a system in which each concept seems to be reached only through a door of unreality where anthropocentricism associates in self-congratulatory smugness with death. Even one of the leading bird organizations refers to grouse, pheasants and partridge as 'game' birds. An American hunting site talks of 'winged targets', reminiscent of Ernest Hemingway's disordered view of pigeons as no more than things to be shot.

I can buy hare too, whole or jointed. Recently, a study of mountain hare populations in the eastern Scottish mountains showed a decline to 1 per cent of their 1950s levels, much of it related to the management of land for shooting and relentless over-culling for the benefit of 'sport'.

Sometimes, as I did that winter morning, I see the people who take part in these things passing by as I walk in the hills or go about the city. More often, I see them in some Deeside town, at Highland games and gatherings, men in plus fours, in camouflage jackets, military sweaters, tweeds, the ones who run the shoots or own them, the ones who work the estates, tend the moors, the 'Guns' who do the shooting.

Their appearance mirrors the language of the books and magazines associated with their sport, often using the terminology of warfare. 'It's as if you're going over the top at the Western Front,' the owner of a top-of-the-range hunting rifle says, although of course, it's not. There's 'up the line', where similes of war transform the flight of birds to 'missiles' or 'fighter jets' or 'low level flying tactics', shooting butts to 'foxholes', a heather moor to 'the battlefield at Mons'. If the birds are the enemy, they're 'quarry' too. They may be 'wily', 'brave' or 'cowardly'. They may be 'alien', like pheasants, 'no more than domesticated fowl' or worse, like 'Frenchmen', which is what red-legged partridges are called, according to one purveyor of expensive hunting equipment, both because they originate on the continent and because: 'they prefer to run unlike our own brave little grey or English partridge – which more readily faces the Guns'.

In the years between the wars, a farmer on Deeside wrote of the conflict and anguish he experienced in his dealings with other species. As a farmer he was, he said, one who killed in the course of making a living but he expressed regret and sorrow at the need to do so. He regarded the 'game' shooting which took place in the area in quite another way, criticizing the vulgarity and waste of shoots and what he saw as the personal inadequacies of grouse shooters – 'rather pathetic people'. In his anticipation of the imminent end of organized shooting on Deeside and the return of the land to agriculture, he was wrong. About something else, he wasn't. A day of shooting, he wrote, must require, 'a great atrophy of feeling, an advanced state of decadence'.

'Decadence', hedonism, lack of moderation, self-indulgence. Only the wealthy can take part in the shooting of grouse – only 'zillionaires' I read in one hunting manual, or even, as the writer puts it, 'parvenu zillionaires'. In an article in a Scottish newspaper in support of grouse shooting, I read that it's deplorable to criticize game shooting on grounds of class, something that doesn't happen in Europe or the United States according to the writer, himself the owner of a grouse moor. Shooting and hunting, he says, are not seen there as being reserved for the privileged class. That might be because in Europe and America, with different systems of land ownership, different history and culture, they aren't reserved for the privileged class. That is not to say class isn't involved, that different sorts of hunting aren't undertaken by the wealthy, that the traditions of Ortega don't live on in elite atavistic hunting organizations, in semi-secret hunting societies such as the International Order of St Hubertus and other worldwide fraternities of wealth, privilege and power with ideologies whose malign voices sound behind the firing of every gun. Critics of game shooting are derided as 'urbanites', 'townies', 'bunny huggers', ignorant of 'country ways'. To be 'urban' is to be detached from the world of 'nature', to be sentimental, denatured, in an echo of Lorenz's scorn for 'the urban and domesticated'.

Cities and their way of life may be a world that is disassociated from the countryside but it's the world where the wealthy participants in driven grouse shooting make the money they spend on shooting, on the *stuff* which, contrarily, is so much associated with 'game' shooting. As long ago as 1948, the American ecologist and forester Aldo Leopold wrote of 'an infinity of contraptions' and decried the proliferation of hunting and shooting equipment as a distraction from the appreciation of the outdoors in an essay in his seminal book on ecology *A Sand County Almanac*. There's gamekeepers' stuff: pens and feeders and drinkers and traps: magnum, Larsen, magpie, spring and pigeon, snares too, fox or rabbit. There are quad bikes, decoys, alarm mines, skull-and-crossbone warning signs, lights and cameras and knives. Far more than that, there's the stuff offered by the prestigious gun companies, the guns, ammunition, clothes – the 'fashion' items, the plus fours, breeks and jackets, stockings, shooting capes, boots and caps and gaiters, the position finders, game counters, flasks and knives and walking sticks, the shooting sticks and gunboxes, the gifts to give the host after the shoot is over, the hunting knives of engraved mammoth bone and scrimshaw, the paintings and prints, the silver, bronze and glass, much of it in the shapes and images of the creatures just killed, all fiercely expensive, beautifully made only for this purpose, this exclusive, timeless, exorbitant, other-world stuff.

The cost of hunting doesn't lie only in guns or days of shooting, in the tips given to the people who maintain the shooting grounds at the whim of the participants – there's a cost in the falseness of the use of words, the insidious erosion of meaning and value. In photo line-ups of post-shoot tallying, ranks of grinning men with guns stand behind rows of slaughtered birds and beasts. 'The shot birds are laid out to honour them', 'Respect for the birds is important,' the shooting journals, books and websites say. Words slip their meaning in this sovereign territory of men: a territory where 'honour' and 'respect' are mutable, nature altered to an equation of man versus 'predator' in the reductive simplicities of warfare and judgement. (Those appearing in the photos are men who are, almost without exception, white.) Every creature seen to threaten a bird being raised for shooting is 'vermin', a 'savage', 'ruthless' or 'voracious' predator – a list published by a shooting magazine some time ago included hen harriers, peregrines, goshawks, badgers, polecats, otters, wild cats, pine martens, ospreys, red kites, hedgehogs, golden eagles, ravens as well as most other species of corvid. This would be a perverse way to look at the natural world except that it is an entirely unnatural world, one managed, manipulated, maintained for the purposes of death. Golden eagles, hen harriers, merlins, peregrines, red kites, buzzards, sea eagles are shot or poisoned deliberately in or near shooting grounds. I read of a game-keeper convicted for killing two short-eared owls on a grouse moor in Cumbria. He lured them with an illegal calling device designed to attract birds or animals with the sound of others in distress, then shot and stamped on them. I don't know which device he used. The company in America where he got it offers 'Predators', 'Screaming Banshees', 'Patriots' and 'Shockwaves'. Satellite-tagged hen harriers, their populations dangerously diminished, disappear into the chill air of shooting grounds. Their mating flight is of such spectacular beauty that it has been described as a 'sky dance'. I've watched hen harriers in Orkney flying low and straight in strong and purposeful flight. The friend I was with there had made an exquisite sculpture of a hen harrier 'food pass', the breathtaking moment of exchange when males feed nesting females in flight.

Conversations stay in my mind, expressions of regret, anger, incidents of sadness, the gamekeeper's son who talked about his father's retrospective bitterness at the creatures he'd been made to kill. The friend who shot a rabbit was distraught on discovering her to be a lactating female, the man in old age describing being taught as a young keeper in the West Highlands to climb up to golden eagles' nests to 'put a stone' on the fledglings to kill them; the friend beating at a grouse shoot who watched one of the 'guns' shoot a passing jay.

What 'atrophy of feeling' is involved when a man can boast, as one American hunter (a prominent lawyer) does, of the species he's killed, the elephants, lions, kudu, the 150,000 birds, of how he'd kill what he knew to be the last duck. What atrophy is required to allow one to take a selfie with a dead giraffe, an elephant, a hundred mountain hares? In his collected notebooks, Loren Eiseley muses on our relationship to others. It's not from guilt, he writes, that he steps over an autumn cricket or refuses to kill the last osprey. It is from empathy, from a compassion that he extends beyond himself to all, regardless of who or what they are.

9

The Coat

It was on an afternoon in a street near home when I passed a young woman wearing a long fur coat. I think she was Scandinavian – she was walking towards the school where a lot of the city's foreign communities send their children. The coat was a surprise. You don't see fur coats here now and haven't done for a long time, probably not since the days of the 1970s and 1980s campaigns when adverts with slogans such as: 'I'd rather go naked than wear fur' or 'It takes 40 dumb animals to make a fur coat. But only one to wear it' were everywhere, when naked models held up the bloodied bodies of mink and rabbits for photographers and anti-fur protestors flung red paint at celebrities in fur coats on 5th Avenue. Then, owners of even the most ancient of inherited fur coats became anxious, terrified that if they wore them in public, they would invite criticism or even attack. This coat was neither ancient nor inherited and although faux fur can be quite convincing, this was clearly real. Long and fashionable, it was made from strands of brown and black and white fur stitched elaborately together and could have come from any of the expensive shops and fashion houses that still sell fur. It is easy to forget, but, despite designers very publicly declaring

244

they've stopped using it, fur's still there, a major industry. I did not know whose fur it was, either mink or fox. In a passing moment, it is always difficult to tell.

It felt like a message, something pointed and timely to remind me of the many natures of reality. Although clearly very expensive, the coat looked tacky and garish. It was unnecessary too – the day was autumnal but not cold, no less than 12 degrees. As I walked on through the crisp scatterings of pavement leaves, I thought about the vagaries of fashion, the wax and wane of it, a bit like the fashion for anything, for music, art, food, one of those unfathomable tides which command what we choose to wear or not or how we decorate our houses, clouds which spread their dimming shadows for a while and then move on, the patches of light, the returning darkness.

I hadn't thought about fur for a while, or about my own long, uneasy relationship with it. There was a lot of fur around when I was a child: fur exuding metaphors, teaching quiet, bitter lessons in the semiotics of contradiction. Fur, I learned early, is both death and sex, triumph and subjugation. Everywhere then, a small, pinched fox face might engage you in glassy-eyed confrontation from the end of its empty pelt, tiny claws dangling fore and aft in attitudes of defeat. Coats of dusty musquash (aka muskrat, *Ondatra zibethicus*) turned foggy, snowy Glasgow into a wilderness of shuffling brown figures in zip-up sheepskin boots. The scent of mothballs was in the air and nose and in the awkward, furry embraces of elderly female relatives as they would press me tightly into their strangleholds of Persian lamb. (Had I known then what 'Persian lamb' was, I would have recoiled, and run.) Marilyn Monroe posed, smiling in her white fur coat.

I knew it too because my father, when he was on business trips to Copenhagen, would buy fur for my mother, gifts to satisfy the staunchless narcissism about which I can write only now that she is dead. He bought her coats, stoles, jackets, all made from different animals, mink, seal, ocelot, and I hated them because I knew what they had been and because they seemed to give my mother more power when she already had enough. The Danish fur company used leftover scraps of seal fur to make into little brooches in the shape of seals with bright, dark eyes. My father brought me one and I loved it and kept it in a pencil tin with my collection of skeleton leaves and sea glass.

Even then, there was a sense of ending surrounding fur. Still flourishing into the later years of the 1960s, it seemed to be falling into desuetude, still there but out of sight, diminishing. For a long time, I did not know if it was just my own feeling that it was in the past or if it simply regressed into part of the life I left when I was sixteen. The Edinburgh premises of the furriers, distant relatives of my mother's, across the road from the flat where I lived when I was a student, seemed to grow dustier until it was hard to tell if it was still open. Those were the years when fur companies everywhere were struggling or going bankrupt. Fur just wasn't fashionable any more or perhaps people had begun to be aware in a new way of what humans were doing to the life of the earth. The publication in 1962 of Rachel Carson's book *Silent Spring* had encouraged at least some radical thinking about man's effects on the natural world, and Lynn White's 1966 lecture on the crisis of ecology with its warnings about the effects of human disregard for nature may have had its effect, but it is impossible to know how many people noticed.

For a long time, I almost forgot about fur. I was happy enough to, or at least to believe that in time the fur trade would disappear from the earth, crushed under the righteous opprobrium of at least some inhabitants of a world increasingly disapproving of what was widely regarded as extreme cruelty to animals. The last time I had really noticed it, in New York winters in the 1990s, I would watch purposeful women in long mink coats, trainers and earmuffs marching the streets of Manhattan, looking like New Yorker cartoon versions of themselves. They reminded me of the past and I believed them to be an anachronism. That was in the days before the economic liberalization

246

of China and a new order in the no-longer Soviet countries began to affect the production of luxury goods, before the newly rich began to demand what the newly rich everywhere have always demanded, and still demand, jewellery, watches, cars, haute couture, fur; before the markets responded as they always do, by ensuring that if the rich want to buy them, the goods will be there for them to buy. Slowly, fur reappeared, in magazines, in adverts, models clothed in garments of fur, fur in items in shops, pom-poms and hats and bags and I realized that something had been changing, moving backwards, that it had never really gone away.

Then, someone sent me a financial report produced by the fur industry, drawing my attention to the astonishingly high figures, the ones which told just how much fur was traded, the revenue, the numbers of animals slaughtered annually to provide for it. It was startling. By the early years of the millennium, the producers of the components of luxury goods of all sorts had begun to proliferate and thrive. Traders, dealers, fur farmers, trappers, dyers, tanners, bio-engineers, designers, fashion houses, journalists, models and public relations executives were all given impetus by the prospect of increased revenue, their efforts encouraged and aided by the well-funded organizations that existed for the promotion of fur. All over Europe, China and Russia, new fur farms were being established. The knowledge was disturbing but at the time I was busy with other things and fur and how it was produced slipped into the space in the mind where topics for the future go to smoulder and nag.

And then, fur seemed to wane again. Designers and manufacturers declared they would no longer use it and the anti-fur campaign seemed to be winning and I was reassured, believing briefly that surely this time it was really over. But then I saw the woman and her coat. I knew that what I had seen was either mink or fox, because mink and red and Arctic foxes are the species most commonly used for coats like that. Walking home, I speculated about how many creatures it would have taken to make, and how and where these creatures lived and died. It did not take long to discover – a coat would take 30–70 mink or 10–20 foxes, 30–200 chinchilla, 60–70 sable, 30–40 karakul lambs, 8–12 lynx, 30–40 raccoons, 200–300 squirrels, 30–40 rabbits, 6–10 seals, or for those so inclined 20–30 cats or 15–20 dogs. The fur used would almost certainly have been 'farmed' since 85 per cent of fur used in fashion is. It is produced in 'farms' in Finland, Denmark, Russia and China, Canada, the United States, Taiwan, South Africa and Namibia, Argentina, Chile and Brazil, and in other countries in Europe, and traded through the half-dozen or so major fur auction houses. The landscape of fur farming changes constantly as countries ban it, announce plans to ban it in the future, or having done so extend the period before the ban will come into force. In Britain, fur farming was banned in 2000 in response to public opposition. Some countries prohibit the farming of certain species such as raccoons, while continuing the farming of others. Some ban fur farming and the fur farms move elsewhere.

Once it was only warmth we required from the pelts of animals, when it was cold and there was nothing else, when fur was what it was, before the progression which took it through time from necessity to the highest refinement of luxury. Fur is there in the language of the past, in ermine, in *miniver*, *fitch*, *foynes*, *lettice* and *budge*, in the garments of the past, *surcote* and *houppelande*, fur-lined and often lavish, in the coats and cloaks worn indoors and out, the squirrel slippers and boots and the fur-lined coverlets for the chill climates of the north, in the apparel of both poor and rich, trimmings of fur on velvet, damask and silk, trousseaux of fur, garments valued, kept, traded, bequeathed. Once it was necessary for warmth, but now it's not.

With the resurgence of its popularity at the beginning of the twenty-first century, it was not only the volume of the fur trade that changed, its very nature did too – the buyers, the sellers, the focus, the garments, the language used in promotion and advertising. The models who had appeared in anti-fur adverts were now seen about wearing items they had once abhorred. No longer the Edwardian offerings of 'reliable coats for motor car wear' and 'moth proof fur coats' (with the added bonus of a 'free 7½ foot Chinese leopard skin' *when you ordered just the catalogue*). No longer the restrained 'distinctive and elegant' or even 'hygienic' garments of the twenties and thirties, or the 'glamour' of the fifties, fur had become infused with a new vocabulary of 'sustainability', 'regulation', 'domestication'. No longer the polite, deferential offering of goods, there was now a bald, in-your-face confidence, a passive-aggressive defensiveness at the prospect of possible opposition to the production and promotion of fur. Adam Smith's description of animals as 'unmanufactured commodities' seemed to equate only to an animal in a cage.

It seems that by 2015, around five thousand fur farms in most of the countries of Europe were producing 42 million mink, 3 million fox, 206,000 chinchilla and 155,000 raccoon pelts. China produced 35 million mink, Russia, 1.9 million, and the United States and Canada, some 7.5 million between them. Namibia sent 120,000 karakul pelts for auction in Europe, and Afghanistan, 1 million. The number of dogs and cats killed annually mainly in China has been estimated by several sources at 2 million and although difficult to verify, is unlikely to be fewer. Fur farming is banned in the state of São Paulo but Brazil as a whole exports large numbers of chinchilla pelts. Precise information about production in South American countries is hard to obtain but it seems that chinchilla furs from Argentina, Chile and Mexico are traded through a Canadian company that deals with 70,000 to 100,000 pelts annually. Wild creatures are trapped for their fur too; between 3 and 5 million muskrat, beaver, marten, lynx, squirrel, raccoon, bear, otter, wolf, wolverine, ermine, badger, seal and opossum are trapped annually in North America alone.

Soon after I began trying to find out more about the modern production of fur, somewhere in another world – that weird world of mind-game algorithms, someone, something, decided I needed regular fixes of offerings of fur. In the corner of my screen every morning, a

Barguzin sable coat, reduced, NOW, from £44,364 to £17,746. A bargain! or a long mink coat in bright pink checks with a lurid green collar, a pastel-coloured mink coat 'overprinted with faux Japanese scenes of whimsical birds and bamboo', a puffy bright scarlet fox coat bedecked with pom-poms, or a 'fluffy, cute birdy' made from mink and fox. Did I want a mink phone cover? A 'bag charm'? A 'partially clipped mink fur and laminated leather' coat for a small girl, which the fashion house told me 'whisks us away to the world of pretty things', which it might do, I suppose, for those receptive to unexplained acts of whisking, not averse to gender stereotyping, or who consider £12,000 suitable recompense for the experience. Or for those insensible as to the origins of their offerings.

The many sables used for the Barguzin coat would have come from a fur farm in Russia where highly prized Barguzin sable were introduced as part of a programme of industrial breeding in 1931. Sable, *Martes zibellina*, are of the family of weasels, stoats and martens. They are creatures of the boreal forest, the taiga; in the wild they are strongly territorial, and leave their hunting grounds only when food supplies run out in winter or when they're feeding young. They nest in hollow trees, crevices, climb with ease, dive into snow after prey – small mammals, birds and insects. They catch fish and cache their food supplies of pine seeds and berries for periods of need. Solitary and crepuscular, they have acute senses of smell and hearing, and, alas for them, also beautiful fur of soft and glossy grey and black and white.

The mink coats, 'birdies', 'bag charms' would have come from one of many fur farms in Europe or in China. Mink, like sable, are solitary beasts, semi-aquatic, riparian, their feet partially webbed for diving and swimming, creatures who climb, leap, forage along the lengths and depths of rivers, digging, snuffling in earth, scenting what they find to eat: small mammals, birds, eggs, insects, larvae. They may have several dens for sleeping, resting or caching food, riverbank dens lined with dry leaves and feathers. Thinking of them, it seems impossible to equate their freedom, beauty and autonomy with nothing but their fur.

I have always hated how mink have been turned by their name alone into symbols of nothing more than a visible accompaniment of wealth.

The original owner of the fur that became the dyed scarlet garment would have been a red fox, *Vulpes vulpes*, although Arctic foxes, *Vulpes lagopus*, are farmed too. They're social creatures who run and roam many miles in a day: 'The Arctic fox spends much of its life on the tundra', a book on the natural history of Russia says. 'In winter, it roams far to the north over the polar ice and southwards to the great forests of the taiga.' Tunnellers, master builders, architects, foxes dig large, complex underground castles, many chambered dens with multiple entrances which are often centuries old, the seats of dynasties, the founding places of generations. In Canada, where aerial studies have allowed them to identify where foxes live by the vivid colours of grasses and wild flowers on top of their dens, wildlife biologists describe foxes as 'gardeners', as 'ecosystems engineers'. You don't see many foxes in Aberdeen and mink are even rarer but a few years ago I saw a dead mink on the roadside, its beautiful pelt lit by winter sun, and another, a still presence poised on a riverbank.

After I saw the woman and her coat, I thought of how these creatures move, with such lightness, with such quietness and grace, and re-read Alice Oswald's poem 'Fox' in which her female fox appears as a creature both of light and delicate movement and a powerful presence as she calls in darkness.

W. S. Merwin's 'Vixen' is an evocation of the mystic, symbolic fox, embodiment of time and man's relationship with the natural world:

> . . . even now you are unharmed even now perfect
> as you have always been now when your light paws are
> running
> on the breathless night on the bridge with one end . . .

he wrote in a poem of extraordinary and profound beauty.

Although I tried, I couldn't think of a poem about mink.

In April 2018, the Environmental, Food and Rural Affairs Select Committee of the House of Commons held a session to review transparency in the selling of fur products in the UK. The CEO of the major European fur organization was asked if fur animals are raised in cages. Her reply was: 'In cages, yes, of course. They are raised in cages.' When asked how big the cages are, she replied, 'A single adult fox would have a cage of 0.8 square metres ... It would have a minimum height of 70 centimetres, a minimum width of 75 centimetres and a minimum length of 100 centimetres,' but she was unable, when asked, to estimate the average size of a fox. The average size of a fox is 25–50 centimetres at the shoulder and 62–72 centimetres long. Asked if the animals were ever able to get out to run about, she replied, 'They are kept in cages. They are domesticated. These are animals that are bred and raised in a cage.'

The domestication of other species has been a prolonged process undertaken to reduce 'wild' characteristics, including fear of humans, in order to make animals more amenable to human purposes. Selective breeding over thousands of years has brought about numerous hormonal, morphological, physiological and behavioural changes in selected species, the genetic basis of which remains the subject of intensive interest and study. Possibly the best-known experiment into the domestication of foxes was begun by Dr Dmitri Belyaev in the 1950s at the Institute of Cytology and Genetics in Novosibirsk. Although undertaking such work was dangerous during the hostile years of the USSR's persecution of geneticists, Belyaev was able to pursue his real interest – the genetic basis for domestication – after Stalin's death in 1953. During the study, which he began in 1959, he carried out a rigorous selection of silver foxes for tameness and lack of fear. Belyaev died in 1985 but the work was continued by his colleague Dr Lyudmila Trut. In 1999 she wrote: 'Forty years into our lifelong experiment ... by intensive selective breeding, we have compressed into a few decades an ancient process that originally unfolded over thousands of years.' The work in the genetics of fox domestication has continued, the results published in 2018 illustrating differences in brain pathways of tame and wild foxes. (One result of the 'lifelong experiment' has been the production of foxes sufficiently tame to be sold as pets to the wealthy of the United States in order to fund still further research.)

Despite the fur trade's continuing assertions that fur animals are domesticated, the selective breeding they carry out is limited. While Belyaev and Trut's work involved extensive and prolonged scientific selection, a 2001 report by the European Commission's Scientific Committee on Animal Health and Animal Welfare confirmed that the fur industry selects mainly for fur colour and only in a limited way for behaviour or breeding possibilities. Despite the trade's continuing assertions that fur animals are domesticated, the website of Sojuzpushnina, the Russian fur auction house, makes it clear that they are not: 'although special selective work in order to tame sable hasn't yet been carried out, it is clear, that it is only a matter of time'.

What do these caged, 'domesticated' animals require in order to live? How much of their natural behaviour can they express? In the wild, foxes run and dig and mink run and swim but in fur farms they are unable to do either. Providing a place for mink to swim would be expensive and so the fur industry tries to question the idea of their basic need to do so. Several years ago, the then head of the British Fur Council, Robert Morgan, argued that farmed mink could not be given water in which to swim because they might get cold and die. In 2001, a study carried out by Mason, Cooper and Clareborough from the Department of Zoology in Oxford was published in *Nature* under the title 'Frustrations of fur-farmed mink'. The article suggested that 'the high level of stress experienced by mink denied access to the pool is evidenced by an increase in cortisol levels indistinguishable from that caused by food deprivation.' Among the other studies into fear responses and anxiety levels in fur-farmed mink, I find ones entitled

'Who's Afraid of the Big Bad Glove?' and 'Sleeping Tight or Hiding in Fright' and yet the fur industry continues to claim otherwise. In 2015, Fur Europe said, 'the claim that mink suffer without access to swimming water is based on the assumption that farmed mink are missing something they never experienced.' A fur trade report based on evaluation of trade practices in Denmark by the Animal Ethics Council, an international advisory body with close connections to member governments, agriculture and commerce, has suggested: 'There is no scientific evidence that foxes need to dig in the ground.'

On their website, a small British company selling hats, scarves and gilets of fur, 'imported from Danish sources', say they have been asked about the moral implications of wearing fur. The now-obligatory statement of self-exculpation reassures me that I need not be concerned as to environmental, welfare or ethical questions because the fur industry is 'intensely regulated'. They sell items made from raccoon fur, which is not produced in Denmark but do not say where it is from. They write of the kindness with which fur animals are treated: 'most animals are even given toys to play with!', a reference to 'cage enrichment'. In some fur farms, 'enrichment' may be provided for animals in order to ameliorate the frustration of their circumstances and the boredom that leads to 'stereotypy' – the abnormal, repetitive behaviour of caged animals, and to self-mutilation and cannibalism. In response to confinement, chinchillas may bite their own fur, chew the bars of their cages, 'backflip' – hurl themselves around their cages and show high levels of 'abnormal repetitive behaviour'. 'Enrichment' may be a place to hide, a piece of wood, some straw. When asked how an animal can be active in a cage not much larger than itself, the CEO of Fur Europe told the parliamentary committee: 'There is food. There are enrichments ... For the fox it would be a bone ... for the mink ... it would be a soft plastic tube.' Some fur farms provide 'enrichment' but many do not because it costs more. A Lithuanian fur farmer defends his failure to provide 'enrichment' because, he says, the way he keeps his animals is a 'tradition'. A quotation flickers in and out of my mind, one I cannot quite place. After some bookshelf riffling, I find it. In 1886, in 'Writings on Civil Disobedience and Non-Violence', Tolstoy wrote: 'I sit on a man's back choking him, and making him carry me, and yet assure myself and others that I am very sorry for him and wish to ease his lot by any means possible, except getting off his back.'

The fur trade suggests that it is 'intensely regulated' but the law, as it applies to fur farming and trading, is often vague or opaque. Within Europe, a complex labyrinth of EU regulations, recommendations, codes of practice, strategies, proposals and directives means that few EU laws regarding either fur-farming practice or animal welfare apply to all member states. Individual countries may have laws relating to animal welfare but standards are not uniform and many relate to agricultural practice but not fur farming. There are frequent conflicts of interest when governments who are responsible for animal welfare legislation are among those who benefit the most from fur trade revenues. Welfare standards in some countries are relatively high, in others they are not. Enforcement of existing laws varies markedly from country to country. Inspection standards are not uniform. If the government of one European country proposes introducing restrictions, fur farmers are able to relocate somewhere else. Outside the EU, there are even fewer controls. Standards of animal welfare in many countries are low and many countries either do not have adequate legal protections for fur-farmed animals, or fail to enforce the laws they have. Many laws relate to licensing or are self-regulatory while restrictions on fur farming in a specific country do not suggest that fur trading, buying and selling do not take place there, or that garments or other items made from fur are not sold there. Neither does its presence suggest that a majority of people support it. In the EU, there are restrictions on the importation of seal pelts and the pelts of creatures trapped in countries where steel leg traps are still allowed, and of the pelts of dogs and cats from China (although identifying imported pelt species is difficult). The international trade in furs means that the pelts of animals raised and slaughtered in one country are traded in another through the major fur auction houses in the United States, Canada, China, Finland, Russia and Denmark. The volume of trade in animal pelts between Denmark and China is such that the headquarters of the Danish Fur Breeders Association, Kopenhagen Fur, publishes a Chinese edition of its in-house magazine. In 2019, they collaborated in a live-streaming online 'fur festival' with a well-known and popular Chinese fashion 'influencer'.

In 2009, a European system of fur certification and animal welfare evaluation, 'Welfur', was initiated by the European Fur Breeders Association. To be fully implemented by 2020, it lays down a complex evaluation system of 'principles', 'criteria' and 'measures' to codify the treatment of animals in fur farms. As both a fur trade initiative and a voluntary one, its remit does not address the question of keeping animals in cages. The fur industry claims in self-vindication that the standards are set by scientists but the research upon which Welfur is based is funded by the fur industry. Regulation and the 'evaluations' of Welfur don't necessarily prevent cruelty – leading officials from Norway, Finland and Denmark's fur trade, all countries which claim to have the highest welfare standards, have been accused of, or prosecuted for, cruelty after video evidence showed animals on their fur farms suffering from serious levels of neglect. One of them, a board member of Kopenhagen Fur and prominent figure in the Danish Fur Breeder's Association, was prosecuted in 2011 for animal cruelty.

What is animal cruelty? Beating animals? Using metal implements to handle them? The means by which they're killed? Restricting the food given to female mink to increase breeding success? Or is it the act of keeping animals in too-small cages all their lives, from May until November, from birth until the 'pelting' season?

In another world, a distorted parallel of the one that tried to lure me with pink fur coats, I now find my screen filled with images from fur farms, creatures crammed into small wire cages in dank sheds, mink and foxes self-mutilating, chewing their own fur or limbs, missing ears, missing paws, creatures with suppurating stumps, raw wounds, matted or missing eyes, bleeding in filthy cages, and worse. The images I look at are made by the television companies and newspapers who feature exposés on fur farming carried out in many countries. One from Russia shows thousands of the skinned corpses of mink lying in raw, pink heaps on muddy ground, filthy cages and animals in states of advanced suffering. A report from a Polish fur farm features the testimony of a worker who clandestinely filmed his workplace and told of fellow workers throwing mink against walls, beating them for fun, stamping on them. 'The minks, held in tiny cages, scream and squeal,' he says. There are many reports and images of hugely obese and suffering foxes on Finnish fur farms. Credible reports from every country in which fur farming takes place tell of extreme animal suffering and cruelty. The fur trade dismisses the evidence as fake, made by 'animal rights extremists', even when made by non-partisan sources such as newspapers, reputable documentary makers and television channels. 'Shocking pictures will always play well to the audience and bashing the fur industry is so easy,' the chairman of the International Fur Federation said after the showing of the footage from the Russian fur farm. In my mind now for ever: the silver fox grappled by the neck from his cage with a metal hook, thrown to the ground and stamped on, his neck and head crushed under a heavy boot, the blood that poured from the creature's smashed nose and muzzle as he died.

Sustainable, ecological, environmentally friendly are the words reached for with the greatest enthusiasm by those engaged in the activities most clearly designed to be deleterious to the planet and to society. They are cover-all words behind which many industries continue to do what they have always done and the fur industry is no exception. Fur farming is similar in impact to other animal farming, contributing to air pollution and ammonia emissions, adding to the environmental cost of the transportation of feed and water usage. But unlike other farming, which produces animals for food, fur farming involves the large-scale disposal of animal corpses once their pelts have been removed. Denmark has a system of using the fat from mink as biofuel and the rest of the corpse as fertilizer but that is not the norm. The large-scale disposal of excrement and poor biosecurity can cause the transmission of pathogens such as mink enteritis virus while the animals themselves may be infected by food or water sources as Danish mink were by swine flu contracted by being fed infected pig-lung tissue. Local wildlife and humans may also be affected. In 2007, the disease necropurulent dermatitis affected mink, and keratoconjunctivitis foxes in Finnish fur farms, spreading quickly between farms. In 2016, there was a severe outbreak of Aleutian disease virus – ADV or mink plasmacytosis – in Danish fur-farms, leading to the culling of 700,000 animals. ADV can affect other mustelids, wild creatures, weasels, otters, ferrets and badgers. There is also the environmental cost of the processing of pelts and their transportation around the world.

In 2011, the UK's governing body for the advertising industry, the Advertising Standards Authority, adjudicated on a complaint brought by the Belgian organization Global Action in the Interest of Animals. Their objection was to a magazine advert placed by the European Fur Breeders' Association, the text of which read: 'Why it's eco friendly to wear fur' and 'one of the most ecologically balanced systems in agriculture'. The majority of the complaint was upheld on the grounds of the absolute claim made, which it was decided could not be substantiated.

In 2013, a report published by the Finnish Environment Institute stated: 'Rain and meltwater cause the nitrogen and phosphorous compounds in fur animal faeces to leach into groundwater or via runoff to surface water bodies ... other problems such as the leaching of drugs given to the fur animals may form a risk to the groundwater quality.' In a letter from the chairman of the organization Coalition Clean Baltic, which represents the Baltic countries, to the European

Commission in 2015, one of the sources of 'pollution inputs' contributing to the serious problem of the eutrophication of the Baltic Sea, was stated as being animal husbandry, including fur farms. Canadian sources report similar findings. In the joyously named *Algae World News* in 2015, Jocelyne Rankin, an expert in freshwater ecology at the Nova Scotia Ecology Action Centre, wrote that the blue-green algae affecting a dozen lakes in Nova Scotia was directly linked to the mink-farming operations which 'have been allowed to expand at an alarming rate in this province with little regulatory oversight'. The social action group Council of Canadians suggested much the same. A report entitled 'On Notice for Drinking Water Crisis in Canada' stated that lakes and rivers had been polluted by the 150 mink farms in the province and that current regulations were inadequate for the protection of water systems.

The processing of fur is similarly polluting. In the same year, an article in the *Shanghai Daily* bore a headline: 'Fur and leather industry leads to hair-raising pollution problems' and described the environmental catastrophe which was being experienced in Daying Town in Hebei province, a centre of the fur-processing industry where dried-up watercourses have filled with sludge and once-clear rivers have become heavily polluted by the unregulated discharge of toxic chemicals used by the tanning and dyeing industries. (How much is attributable to leather and how much to fur is not discussed. Leather is already well known for the polluting effects of its processing.) The report suggested that the local government, the body responsible for policing environmental regulations, 'relies on the fur and leather industries for a large share of fiscal revenues'. Environmental degradation was so severe that, according to the *Shanghai Daily*, in spite of the large dowries being amassed by the town's wealthy, 'few outsiders want to marry women from Daying because of the town's battered reputation'.

Many fur organizations suggest that fur is 'natural'. It is as long as it remains on the animal or when it decomposes naturally, organically, after the death of its owner but the moment it is removed from an animal to be subjected to the series of chemical processes which preserve it, it becomes de-natured, artificially preserved to allow it the longevity boasted of by the fur industry (a longevity dependent on environmentally and economically costly annual cold storage to maintain it at an optimum temperature of 55 degrees and 55 degrees humidity and protect it from insect infestation). The dyeing and tanning processes undergone by fur are so extreme that after their implementation, proving which animal any fur came from can be difficult because, according to mammal expert Judith Eger from the Royal Ontario Museum, 'once the fur has been treated, genetic sequences are virtually destroyed'.

Studies demonstrate the hazards not only to the environment, but to the processors, sellers and wearers of garments made of fur, as well as those who live in the vicinity of dyeing and tanning operations. Many studies into pulmonary function in workers in fur processing have shown significant, deleterious health effects. The results of one study entitled 'Poison in Furs' were published in 2011. Undertaken at the Bremer Environmental Institute at the University of Bremen on behalf of EcoAid and Vier Pfoten Animal Protection Foundation, they illuminate in some detail the toxic substances found in fur. The foreword written by Dr Hermann Kruse of the Institute for Toxicology and Pharmacology at the University of Kiel mentions only a few – chromium salts and high levels of formaldehyde, harmful to the respiratory system, alkylphenol ethoxylates, harmful to the hormone system, chlorophenols, 'known to cause kidney and liver damage as well as skin lesions' and aromatic amines, 'mentioned because of their high carcinogenic potential'. These findings have been replicated in other studies and individual reports on product after product made of, or decorated with, fur – most manufactured by the best-known high fashion brands and many designed for children – end with the words: 'Warning, should not be sold, bought or used.'

On the website of Fur Europe, I read a large headline: 'Animals killed annually for: Fur 70 million, Food 150 billion: Morality is not a luxury'. 'Morality', Fur Europe says, 'is not a "help yourself" luxury buffet table from which people can pick and choose ... The very purpose of western philosophical systems is to establish a consistent moral framework from which to make moral decisions.' Fur Europe argues that if we continue to eat meat, eggs, yoghurt or wear leather or wool while considering fur 'an unnecessary luxury good', we are both morally inconsistent, and 'ethical subjectivists', that is, we adhere to the vacuous and indefensible philosophy which bases its moral positions randomly, without recourse to underlying moral good or otherwise. (On the day when I consider I require moral instruction, I will not seek it from the fur trade.) The number of animals killed for food makes no difference to the numbers killed for fur. One number neither explains nor justifies the other. Many of us either do not wear or eat animal products, or we exercise strict criteria and judgement about the places from which we purchase them and the means of their production and it is entirely consistent that we still object to fur. We may be carnivorous, vegan or vegetarian. We may adhere to the moral codes and tenets of our own religions, traditions and personal beliefs in relation to the treatment of animals and the natural world. Many who dislike the fur trade dislike the meat trade too. Food, whether or not one eats a particular foodstuff, has the notable virtue of being necessary which clothing made of fur is not. As Fur Europe suggests, fur may well be labelled as 'an unnecessary luxury good' which is because it is an unnecessary luxury good. You could argue about your right to own these things, about the 'hard work' that enabled you to buy them, about your freedom of choice and purchase but you still could not claim that they are necessary. The wealth of those who can afford to buy furs is likewise immaterial. There are many people who can afford to, but still do not buy fur.

Concerns about the treatment of fur animals are regarded by the fur trade as no more than sentimentality, encouraged by the influence of the Disney corporation. 'We all grew up with wonderful stories of Mama Bear and Baby Bear and we all love Bambi. But nature is not Disneyland,' says the Fur Council of Canada. The chairman of the International Fur Federation agrees: 'Because mink and sable are "cute" we get lost in Disney thinking.' The Danish Fur Commission says: 'The members of the Council dissociate themselves from the so-called "Disney-fying", meaning that the entertainment business accredit farm animals human qualities.'

There is no shortage of abject sentimentality in human attitudes towards other species but to characterize opposition to fur farming this way is to demean the entire structure of questioning how humans treat other species and human responsibility towards them. There may be some whose views are sentimental but there are many others who appreciate the nature of nature, who acknowledge the darkness and death at the heart of all natural processes, who look with equanimity on the outcomes of predation in the necessity for the continuation of life but who still regard the 'farming' of animals for their fur as egregious, destructive, purposelessly and wantonly cruel.

And then, there is the morning when my online offerings *du jour* are a silver fox and 'Orylag' hooded coat and a jacket made from 'broadtail'. I don't recognize the name 'Orylag' so I look it up. The French website promoting whatever it is, is clearly devoted to the panegyric form – ah, the qualities of the product! The modernity! The colours! The lightness! Like silk! Like precious jewels! Irreprehensible! Ecological! Part of the food chain! The product can be dyed to resemble leopard, tiger, chinchilla! It can be decorated, figured, embellished, woven! Whatever 'the product' is, it sounds deeply mysterious, newly invented, like graphene, except it's not. I thought the letters 'l', 'a' and 'g' might be significant, as they are. Lagomorph. Of the order *Lagomorpha*. Rabbit.

I have always had considerable regard for rabbits – any I have known have been intelligent, wilful, independent, possessed of a self-certainty, which always seems to magnify the cruelty bestowed on them by the fates or by biology in the dubious gift of fecundity, the very element which humans appear to believe allows them to be treated as they are. The large number used annually worldwide in animal research pales beside the number killed for fur each year. *A billion.* I can't even find a way to visualize a billion. Orylag is a French trademark, the product of research at INRA, the French National Institute of Agricultural Research, to whom the product is now patented, at their experimental farm at Magneraud in the Charente-Maritime. 'The product' is the fur of a Rex rabbit 'genetically altered to reduce thicker hairs, making their fur particularly soft and smooth and therefore conferring high added value', according to an INRA paper entitled 'Characterization of the gene and mutation responsible for the rex hair trait in the rabbit'. Sixty per cent of production goes to the fur trade, the remainder is sold to the meat trade as 'Rex du Poitou'.

'Irreprehensible' is a matter for discussion. Orylag rabbits are kept in the much same way as other rabbits raised for the fur trade. They are caged, isolated, subject to the physical and mental harm associated with permanent confinement, being artificially inseminated and bred too frequently before being transported for slaughter in crates and killed, in the case of Orylag rabbits, at twenty weeks, by having their throats cut. Rabbits are often slaughtered with great cruelty. Reports of these usually rather quiet creatures screaming in pain as they are killed are common. The word 'rabbit' itself seems to inspire triviality, an inheritance possibly from the language and iconography of rabbits in childhood literature. Many items manufactured from rabbit fur are silly, disposable, whimsical; pom-poms and bobble hats, key rings, 'bag charms', multicoloured backpacks of infantile design. Many people buying these items do not realize that they are made from real fur because they cost so little. One company sells a long 'ethically sourced' rabbit scarf. Some are too fastidious to mention the word. The British gun company that sells vastly expensive items of clothing to the hunting fraternity offers a scarf with fur down the side, which it describes as being made from 'lapin'. An Italian luxury company sells a very small toy rabbit as a gift for a child. Called 'My Little Bunny' it's 'artisanally crafted', costs £1,500 and is made from Orylag. It may be that, despite their circumstances, the vast number of rabbits killed in European fur farms are more fortunate than their millions of counterparts in China. Most rabbit fur comes from China where welfare standards are, at best, fluid. Many items sold as artificial fur are, in fact, Chinese rabbit fur. Recent newspaper reports on rabbit farming in China were accompanied by images of breathtaking violence. Angora, the wool derived from rabbit fur, is obtained by ripping their skins away.

Even beyond the cruelty of this trade, there is something about the making of tens of rabbits into a bright orange garment decorated with pom-poms which seems a step beyond, a step into disrespect, into violation, which makes turning the pelts of many foxes into a coat shaped like a large, puffy red heart into a mockery of human creativity and of the lives of animals. I think again of Lynn White: 'We are not, in our hearts, part of the natural process. We are superior to nature, contemptuous of it, willing to use it for our slightest whim.'

The themes of sex and death are never distant. 'The globally conscious young woman, sensational in sable, mischievous in mink, fierce in fox – represents the epitome of responsible chic' a fur trade advert says. In everything connected with fur, the exercise of money and power burns with a low, malevolent flame. It is not only in human power over other species, it is in those enduring contradictions, the over-sexualized 'glamour', the exploitative images, the atavistic misogyny, the easy disparagements of age: 'Not your grandmother's fur coat', 'in one shot, wearing nothing but the coat. Kate Moss still looking good at 42.' It is in the fur trade's assiduous wooing of 'celebrities', the aiming of campaigns specifically at young women, the use of 'influencers', their jejune and gushing fur 'blogs'. It is in the offers of scholarships, internships, bursaries, courses and competitions for young designers, in the photos of industry gatherings of mostly men, fully dressed in elegant suits set against the images of near-naked female models in fur. 'Morality,' the fur trade says, 'is not a luxury', but it may be when economic prospects are circumscribed for the young, when house purchase and even rent are beyond reach for many of the working young, or when, for luxury companies, there is just so much money to be made. In one industry photograph, a group of young designers, recipients of industry awards, pose happily, all of them clothed in long chinchilla coats.

On the morning I'm offered Orylag, I am offered 'broadtail' too. This is another name for astrakhan, for Persian lamb, for the trade name 'Swakara' or South West African karakul because it is the fur of the karakul sheep and much of it comes from Namibia, formerly South West Africa. It comes from Kazakhstan too, and Afghanistan. Namibia exported 120,000 karakul pelts in 2015, and Afghanistan 1 million. Kopenhagen Fur auction records show karakul pelts being bought by the most expensive European fashion houses. Its website displays garments made of karakul. 'Stunning Swakara' it says, over a display of models barely clothed in scanty items made from karakul fur.

Another word for this fur, one not used much in fur trade advertising is 'foetal'. Karakul fur is most valuable when the curls are still neat and close and tight which they are before the lamb is born. Karakul lambs are often slaughtered at, or before birth, after they have been removed from their slaughtered mother. A karakul sheep typically is allowed to give birth to three lambs before being slaughtered with her fourth, two weeks before term. Animal protection laws in Afghanistan might not be to the forefront of public thinking, although there are plenty of people concerned about the fates of other species. In the light of the country's history and sufferings, it is difficult to separate one cruelty from another, one urgent necessity from another. In 2015, the *New*

York Times reported that, when in power in Afghanistan, the Taliban wanted to ban the karakul trade on the grounds of its extraordinary cruelty. Among the Dead Sea Scrolls a commentary was found: 'And an ox or a lamb with its offspring, you may not slaughter at the same time. And you shall not strike the mother with her offspring.'

One difficulty facing fur farmers is how to kill an animal without damaging his or her fur. As the sole aim of the endeavour is to produce quality pelts, the choice of killing methods is limited. The choice of words is too. In the fur trade, when the subject of death is mentioned at all, the word on every website, in every publication, is 'euthanasia'. In this trade, there is no killing, no slaughter. There is only 'euthanasia'. In this trade, the time for killing is 'the pelting season', death is 'euthanasia'. The ways we use words for the death of both humans and animals are unique in their powers of softening euphemism, in the ways we file down the sharpened edges, ease our minds through smooth tunnels of eviscerated meaning into the eternal darkness of mortality. We 'pass'. We 'pass on', 'pass away', 'pass over'. We 'go beyond', 'slip away' and carry out these unlikely physical and linguistic manoeuvres to lessen the impact of what we know and fear. Animals are 'put down', 'put out of their misery'. Only the meat trade uses the honest word 'slaughter'.

Euthanasia, 'a good death', is the act of assisting a person or animal to die in order to end unbearable suffering, the meaning of this word fixed in almost every context and sphere other than the fur trade. The concept of human euthanasia, killing as the merciful ending of suffering, has been and remains a long-standing topic of philosophical and religious debate. Opposed in its principle by Pythagoreans, stout believers in the transmigration of souls, it was accepted by the Stoics for whom death was a subject to be contemplated daily, with the aim of overcoming the fear of an inevitable event.

There is nothing merciful about the deaths of animals killed for their fur, except perhaps in the curtailment of their suffering. There are still places where traps are used to kill animals, although the legislation allowing or preventing their use in different countries varies. (In practice, enforcement is difficult.) There are three common types characterized as 'killing', 'drowning' or 'restraining'. There are steel-jaw leg traps, 'kill' traps such as the Conibear, which, in spite of the fact that it is designed to kill outright, often does not, and there are snares. The steel-jaw leg trap, although internationally regarded as the most cruel and consequently banned in many countries, is still widely used. With almost admirable frankness, the Turkish Ministry of Agriculture, as part of a rather opaque web page about fur farming, suggests that: 'Since furs with bullet hole are of poor quality, use of trap or insecticide is recommended for hunting.' (sic) *Hunting*.

After a life in a cage, the only thing left for fur animals is the 'pelting season'. This is when, after their first moult, foxes are hauled from their cages by means of metal neck tongs and, unsedated, electrocuted with 110 volts administered through electrodes applied to their mouths and rectums by people who may not be required to have training in carrying out the procedure. Mink are killed by gassing. The words I read about fur-trade practices ring with history: 'mobile gassing units', 'unfiltered exhaust gases', 'killing chambers'. The resonances reach back into darkness.

In 1939, in a commodious villa at Tiergartenstrasse 4 in Berlin, the state sanctioned the murder of disabled children, a procedure that had been taking place clandestinely over the previous years in centres all over Germany, and was now to include the murder of adults. These acts were the practising ground for the future, for the killings, which began in the 'gas vans' of Chelmno, Belzec and Sobibor, and extended to the mass killing operations in which millions died. The name given to the Tiergartenstrasse programme was 'T4', the word used to describe these murders 'euthanasia'.

A brilliant autumn, days of chill sun interspersed with days of chill grey, illuminated by the turning of green to bronze to gold. I stop work for a while and as a distraction I go into the garden to chase and rake the thick layer of leaves which has descended from next door's beech to lie over the entire garden. The leaf pile fills twenty-four sacks. I fill them in stages, carry them through the house and load the car. I clean out the pond, which has a slow leak. It'll have to be replaced and so I plan a new one. The frogs are long gone and I'd be delighted if they chose to return. I want to replant and create a new wildlife garden. I want to think of something else but I go back to my desk and read more about the effects of gassing on other living beings.

The gases used to kill mink are carbon dioxide and carbon monoxide, the latter sometimes supplied by the exhausts of petrol engines. Despite legal requirements, gases are not always filtered or cooled. Mink are taken from their cages and placed in numbers, sometimes up to a hundred, in 'killing boxes'. Often, they die from suffocation and not from the effects of gas. Adapted for water, mink react to low levels of oxygen and often take a long time to die after prolonged distress and convulsions. A Finnish study, which questioned fur farmers across Europe, reported that some simply wait until any noise has stopped to ascertain that the animals are dead, some check 'the state of the animals' through an inspection window and fewer than half check each individual mink for breathing and movement.

In writing this, I write of humans and of other species. There is no way of measuring equivalence or difference. With the damage we inflict, it is ourselves we damage, the lives of us all that we lower and ruin. The differences between species are less important than our humanity and all humanity is tainted by these actions. When we lose our humanity, we do it in increments, in words, in ways, in small and distant actions until they are small and distant no more. I think of Leonard Woolf and the moral generosity of his extension of the universal 'I' beyond the boundaries of his own species, of Tom Regan and his 'biographical creatures'.

Despite publicity given to fashion houses that no longer use fur, the trade persists and thrives. Fur is still produced in vast quantities and sold both by large and small companies and by otherwise reputable shops. In 2019, one of Britain's best-known 'high-end' shops was found to be selling a coat made of lynx fur from an unverifiable source, although it seems that the animals were either trapped or farmed in Montana, and produced by an Italian fur company. Most lynx species are critically endangered. None are 'domesticated'. The shadows move and flit and fall and are very little different from the past.

It turns out, there is a poem about mink, written by the American poet Toi Derricotte who in 'The Minks' describes her uncle's mink farm, his looming, omnipotent presence, the five hundred steel cages of imprisoned animals. She writes of her own timid sympathy for their lives of confinement, how she'd call out a tentative greeting to each as she gave them food and of her sense of there being a wider meaning in the beauty of the fur to which these animals were reduced.

It grows cold suddenly, everything frosted into stillness for a few days. I talk to a new neighbour who gets up early for work and tells me that he has seen a fox in the secluded area behind our gardens and now I know what happened to my doves.

I work and huddle at my desk and then it's mild again. I read wonderful poetry and think about the way we treat other species and one day in the street I pass a woman in a long fur coat.

10

What is Love?

The days of heat of summer will be over soon, the long, lit days when small birds lie in rows, flattened, wing-spread, cooling themselves on the log-store roof. The fledglings in the ivy will leave the garden, as will the speculating magpies and jackdaws whose seasonal food they are. The swifts will go, the days will change and the air. After the summer solstice, the year gains on us and we change too, almost overtaken by the season. In the dove house, the perches are still there, the nest boxes, the food containers, empty. Over the months, there hasn't been trace of the fox anywhere, in the garden or beyond. Every now and again, I find that the leaves beside the pond have been disturbed but a few scattered, decomposing grey and white feathers are too sparse to tell anybody's story.

My last dove never came back. I went out to feed her one morning but she wasn't there. I checked all day but the only birds who were there were her wild companions. In the evening, I swung the beam of my torch into the corners of her empty house although by then I knew she wouldn't be back. The signs had been there in her flying off in a flock with other birds and staying away for nights and days. If I've left her house more or less exactly as it was, it may be from indolence, not sentiment.

Since she left, I've thought about what she meant or symbolized and is, or was. I miss her as I miss them all, at some moments completely and at others partially. She was the final link connecting me to all the other doves and to the years I kept them, to how they marked and characterized my life. Change is always disquieting. The things I did for them have left a space, the feeding, the cleaning, the occasional unheroic rescue.

They were always free to stay or leave. All that impelled them was their homing instinct and understandable response to fear. I'm free too now in a way, all the knots of willing obligation that tied me to them for so long loosened by this last departure. Still, something of her stays, a thin thread of wondering anxiety, the kind you always feel about those lost or flown. I wonder where she'll go. I think of my many generations of beautiful, independent birds. I came by them by chance. I was given two, kept and nurtured them. Their numbers grew and fell and this bird was the last of them.

Doves may seem unlikely creatures to love but I did, for themselves and because they connected me with other worlds and ways of being but they were neither 'pets' nor 'companion animals', creatures who by choice we remove and protect from the rigours of the world. Our arrangement was mutual – I fed and housed them for the pleasures of their presence and in return they stayed, a relationship more commensal than symbiotic. I accepted I was outside their intimate and dove-ish circle, happy to extend devoted, long-term, asymmetric love. It was different from the way I love my rook and crow, both members of species with a rich history and culture of close social relationships. I don't know if the love I feel for them is returned but that is only as it is with any other love.

Our love for other species is ancient, even more ancient than that for the puppy at 'Ain Mallaha, together with her mistress in death

13,000 years ago, or the carefully looked-after Bonn-Oberkassel dog 1,000 years earlier, neither the first nor the only representatives of what might be this often deep, sometimes mutual, appreciative love; this possibly simple, possibly complicated, expansive, generous or quiet, undemonstrative, private *tendresse* we may feel towards other beings. From the first-caught bird or domesticated creature, the relationships we've forged have been significant, twining our lives together through the cultures of the centuries.

Evidence of closeness to other species from the ancient world is often a post-mortem affair, expressed in epitaph and eulogy: 'My eyes were wet with tears, our little dog ... never again shall you give me a thousand kisses ...' is inscribed on the grave of one Roman dog. 'Tears fell for me and the dust was heaped above by my master's hand' on another. Catullus's well-known poems, written after the death of his lover's pet sparrow are typical: 'Lesbia's sparrow ... has now hopped alone down that dark pathway from which there's no return.' (Given Catullus's reputation for salaciousness, some poems are suspected of being about something other than a bird.) Ovid's lament 'On the Death of Corinna's Parrot' is a full-on encomium on the parrot's life and habits, describing the small tombstone on which is written: 'His mistress's darling'. Affection for other species was marked – goldfish, apes, cats, weasels, dogs, hares, ducks, geese, snakes and many other creatures were kept as house pets, subsequently portrayed and memorialized in every medium, a bond in word and image uniting us across time.

Some humans at least must always have thought of themselves as 'animal lovers', as many think of themselves now. We like to believe our countries are 'nations of animal lovers' but if the words 'animal lover' are fraught with ambiguities, so are manifestations of human love for other creatures. Our love often seems a strange and cross-hatched thing, a Venn diagram gone crazy, an explosion of ordered subsets blown to the sky. 'Love' is the word frequently used for anything but love.

The boundaries we set for how and who we love are mutable and capricious: this one is to be the beneficiary, that one not. This one is to be specially chosen, bought, nurtured, recipient of our apparently boundless but often unreliable affection; that one is to be industrially bred, slaughtered, chased, hunted, shot or culled. Like ripples circling outwards from the clear and certain centre of our ourselves, our fickle likes and dislikes have, in one way or another, condemned, judged and sentenced every species, wild, urban, rural, large or small, on earth. Despite our vastly increasing knowledge of consciousness in other species, proofs of their high cognitive functioning, evidence of their sentience and emotions, their capacity for moral behaviour and centrality in the maintenance and functioning of every ecosystem, our feelings for them still seem rooted immovably in past ideas, in human exceptionalism and the speciesism of which Richard Ryder wrote, which justifies our treatment of them on the grounds of their species alone. Persistent anthropomorphism puts human interests first, frequently attributing the worst aspects of humanity to other species while denying them the possibility of the best. The first critic of this way of thinking, Xenophanes, an itinerant, pre-Socratic philosopher and theologian, was harsh about both Hesiod and Homer for portraying gods whose indulgence in 'theft, adultery and mutual deceit', was as egregious as any humans', commenting scathingly that were horses and cattle able to draw, they would draw their deities in the forms of horses and oxen. Even today, we appear to portray and judge every species against the measure of ourselves.

Our decisions about which species we favour and why seem increasingly influenced by the commercialization which has overwhelmed us all. Neither Adam Smith writing of domesticated animals as 'unmanufactured commodities' in the eighteenth century nor Max Horkheimer formulating his skyscraper-rising-over-an-abattoir metaphor for capitalism in the twentieth could have anticipated the turning of animals into both goods and consumers that has taken place in the centuries and decades since; in the vast food, clothing and 'pet' industries, in research, hunting, illegal poaching and 'wildlife tourism', all of which use other species to fulfil every human want. If other species have become products, we have become their complicit consumers.

There is a multiplicity of channels of persuasion and information, from heavily sentimentalized films and cartoons, spectacular wildlife documentaries and omnipresent advertising to the online information exchanges that bombard us from every forum. They have all influenced how we see and treat the animals we claim to love. But where in their transformation from natural beings to consumable goods is evidence of love? Is it shown by people's desire to swim with bottlenose dolphins in the Bay of Islands of New Zealand whose populations are described by their Department of Conservation as 'struggling' because tourists are 'loving them too much'? Is it evinced by the taking of 'selfies', which endanger and sometimes even kill the vulnerable young of endangered, solitary forest creatures, or wild savannah creatures we cross half the world to see? Is it in exploitative captive breeding or breeding-for-hunting programmes? In our desire to buy and own illegally captured and traded wild and 'exotic' creatures whose photos we see online, the otters and cheetahs, sloths, and slow lorises, the amphibians, reptiles, fish, the bulbuls, mynahs and parrot species we're driving to extinction?

How does love relate to the commercialization which encourages us to see everything in terms of acquisition, that allows us to regard animals as *things*, objects to be ranked, selected, bought, sold, handled or disturbed without the care we accord ourselves? Despite a marked increase in the belief that animals have a claim to humane treatment, legal protections and rights, rigorous analysis of who we all are as either humans or not and how we relate to one another, seems still little more than an academic pursuit which seldom strays into the purlieus of ordinary life. We still divide the living world according to the strict discontinuities between *us* and *them*. The concepts of 'personhood' by which animals are considered individuals and thus entitled to the respect we give to humans, or that of 'animality', which attempts to find common ground among all living species beyond the narrow definitions of 'human' and 'animal', are still remote when applied to our everyday dealings with others. The ancient religious–philosophical arguments about human supremacy on which our lives and economies are founded seem as entrenched as they ever were, as damaging and expedient. The delineations between 'human' and 'animal' are fixed and stark. Granting the right to an unhindered life, privacy and freedom from human proximity and a right to the preservation of habitat to other species still appear to be regarded as unwarranted restriction on our 'rights' as humans to do exactly as we choose.

Watching a lizard on his terrace, the American poet Theodore Roethke mused about who the place belonged to, acknowledging that the 'limestone crumbling into fine grey dust' and the worn garden chairs in fact belonged to the lizard and to forms of life older than his own. He even wondered if lighting a cigarette might disturb the lizard. How many people would share his view or sensitivity? How many people believe it important to consider what other creatures think or feel, or to respect their entitlements to their own lives, security and environment? Who thinks about how other species see the world?

A socially conservative critic of Darwinian evolution seems an unlikely person to have put forward the view of other species that would influence many of the great European philosophers of the twentieth century. In a remarkable monograph published in 1934 'A Stroll Through the Worlds of Animals and Men: A Picture Book of Invisible Worlds', Jakob von Uexküll, an Estonian-German biologist, presented the idea of the 'Umwelt', the particular, individual sensory world in which he believed each creature lives. Escorting his readers through a flowery, insect and butterfly-filled meadow, von Uexküll invokes the image of a soap bubble surrounding each creature, enclosing a unique perception-environment he describes as the 'self-world of the animal'. Deeply critical of the fashionable behaviourism of the time, which he believed reduced creatures and humans to 'mere objects', von Uexküll explores the richness and variety of the 'Umwelten' of small creatures such as tick, bee, house-fly, spider, sea-urchin and even those of larger ones – astronomer, deep-sea researcher, chemist and nuclear physicist among them, the latter existing, he believed, in an Umwelt of 'a mad rush of infinitesimal particles'. In its challenge to the human-dominated view that we – who and whatever we are – share one common set of perceptions, it presents the case that 'every living creature is a subject that lives in its own world, of which it is the centre'. Writing before much was known about their neural capacities, von Uexküll acknowledged the complexities of other species' sensory worlds.

It was unsurprising that the Umwelt became a focus of thought for those philosophers pondering the nature and relationship of animal and man. The concept was of particular interest to the phenomenologists Martin Heidegger and Maurice Merleau-Ponty who were engaged in the study of consciousness, subjectivity and experience, and later to Gilles Deleuze and Félix Guattari, exploring the fluidity of human–animal separations.

When the French philosopher Jacques Derrida famously found himself being stared at by his pet cat while naked, the experience of feeling embarrassed in her presence inspired the lecture series given in 1997, later published as the book *The Animal That Therefore I Am*, an extended deliberation on the relationship between humans and animals. The cat was not, Derrida assures us, just any old philosopher's notional cat who only theoretically gazed upon his nakedness, causing him to think at length about the ideas of Heidegger and other philosophers, and delve into the theology, ethics and logic of the divisions between human and animal. It was his own cat, 'a real cat, believe me, a little cat', of whom he was clearly fond. Fascinating, witty and deeply sympathetic, the book is also a discourse on the violence perpetrated against other species and our own in the past two centuries, described by Derrida as 'industrial, mechanical, chemical, hormonal and genetic', carried out for what he calls 'the putative human well-being of man'. He poses questions about responsibility and the ethics of human behaviour, suggesting that it is animals themselves in juxtaposition to humans who provide a means for our self-understanding. The Italian philosopher and animal-rights advocate Leonardo Caffo believes that the encounter between Derrida and his cat marks the significant moment when 'philosophy recognised the animal as a unique, unrepeatable individual', in being seen as capable of subjective view and thought.

Derrida and his little cat are part of a long tradition, successors to the line of those involved in the ancient relationship between humans, cats and other creatures and of the love, appreciation or even embarrassment that have inspired so much memorable human creativity. Christopher Smart, the clever, troubled eighteenth-century poet, wrote a long and marvellous poem, 'Jubilate Agno', during his incarceration in St Luke's Hospital for Lunatics in London, the best-known part of which constitutes seventy-five lines about his cat:

> For I will consider my Cat Jeoffry
> For he is the servant of the Living God duly and daily
>   serving him.
> . . .
> For in his morning orisons he loves the sun and the sun
>   loves him.
> For he is of the tribe of Tiger.
> . . .
> For he is a mixture of gravity and waggery.
> For he knows that God is His Saviour.

277

Incantatory and ecstatic, the poem is both tribute to the cat and praise poem to God, an expression of love, as well as demonstration of Smart's acute observation of feline behaviour.

A similar touching closeness is expressed in the ninth-century poem 'Pangur Bán'. Written in Old Irish, it describes the relationship between a scholar and his cat, both diligently at work on their own pursuits, the cat hunting, the scholar pondering his books. In Seamus Heaney's graceful translation the two of them are: 'Adepts, equals, cat and clerk' in a lovely evocation of a quiet, gentle way of life shared by members of different species.

Our love extends narrowly. Cats and dogs are the recipients of the majority of our other-species love. Both are the most frequently chosen as 'pets', 'companion animals', often – although not invariably – cosseted, loved, protected. Trade statistics suggest that 70 million European households and 84.9 million American ones now keep pets – birds, small mammals, reptiles, fish and, predominantly, cats and dogs who may be favoured because history has brought them to us along shared pathways of geography and time. Both have adapted and evolved from their origins in wildcat and wolf, spread across continents, domesticating over thousands of years, living with us commensally, our familiar, if once-wary confrères and co-workers in the business of the days. Useful and sturdy, they have been our guards, our pest controllers, companions in hunting, as well as representatives of other possibilities and relationships, growing ever closer to the centre of our lives.

I must have been about four when we drove en famille to buy a dog. The day is now only a haze of Sunday afternoon impressions of rain and green, of the muddy track somewhere in the Stirlingshire countryside, a room, a log fire, the two chosen puppies who would be the confidants of my growing up. The black dog died when I was in my early teens, the brown one, the last dog I knew well, shortly before I left school. Our buying them must have been part of the growing tendency to post-war pet-keeping, which had been increasing since Victorian times and was about to expand into the vast pet trade of today.

But what makes us choose one creature over another? Many studies have evaluated the importance of a species' appearance in determining its popularity, commercial potential or conservation status. The conclusions are dismaying: 'An animal's attractiveness substantially increases support for its protection,' one study says while another concludes: 'A few charismatic and cute species ... tend to receive most of the conservation funds and policy attention.' Creatures are ranked: 'the twenty most charismatic species' or described as 'powerful commercial icons' or 'the world's cutest animals'. Even the birds in our gardens are subject to our caprices. The results of a study on the 'likeability' of garden birds show that we like 'songbirds' (even though we may not be able to define correctly what a 'songbird' is) preferring robins and blackbirds to corvids, gulls, pigeons and starlings. We consider the former attractive but the latter argumentative, competitive and noisy, all necessary, natural behaviours of wild birds. '*Charismatic*', '*iconic*', '*cute*' – in a time of devastating and irreversible species loss, can these really be the measures of our love?

What about invertebrates? In any measure of love, we do not include thoughts of ecological niches, of trophic cascades, of the unseen, unknown benefits which we gain from other species in ways we might not understand. Species we regard as malign, the ubiquitous Highland midge or winter moths may be problematic simply because they are inimical to the interests of humans. Parasitoid wasps are efficient controllers of common garden pests. Parasitiformes and Acariformes, the mites and ticks, all million species of them – most as yet undescribed – have important and complex roles in ecology but fall very far outside the boundaries of our interest or concern. Earthworms, who we tolerate because we know of their benefits to our gardens, are never likely to be regarded as 'charismatic' species, but Darwin himself in his monograph 'The Formation of Vegetable Mould Through the Action of Worms With Observations on Their Habits' writes with enthusiasm and even warmth of his discoveries of their likes and dislikes, of their intelligence and their unexpected abilities. That the vast majority of the world's species, 65 per cent of them invertebrate, will fail all the common tests and judgements we construct, belies their overwhelming importance and significance. 'In terms of life continuing on a healthy planet,' the organization Buglife states, 'invertebrates are far more important than we are.'

In an astute and moving essay, 'Praise Song for the Unloved Animals', the American writer Margaret Renki pays tribute to the role played in the complex systems of life and renewal by some of the reviled creatures of the earth, among them opossum, vulture, spider, wasp, bat and snake, and their place in the cardinal cycles of consumption and continuity. She writes beautifully of the red bat's 'canny wings', of mosquitos providing food for the 'chittering chimney swifts' and of the 'glossy vulture' – often mistaken in flight for eagles, ospreys or hawks, 'creatures we thoughtlessly love much more' whose eating of the bodies of dead creatures is a vital stage in the process of returning flesh to life. She reminds us that perhaps, had our love been different, the world might have been too.

If an emphasis on appearance has had vastly damaging effects on all species, it has exercised a cruelly malign influence over those we keep as pets. Once bred for their qualities as working or hunting animals, for speed and strength, the 'selective' breeding of dogs over centuries created diverse breeds from the single canine line but in more recent years criteria for selection have changed in response to the demand for 'pedigree' animals who conform to particular standards of behaviour and appearance. For more than dogs alone, the way a creature looks seems a major determinant of their fate. Beginning with an already narrow gene pool, selective breeding has greatly increased the incidence of disease in these animals, many of whom, as a result of our choices, suffer from life-limiting or chronic, painful conditions.

I stand at the traffic lights waiting to cross. A young man beside me holds a lead – at the end of it is a puppy who stands patiently between us. In the moments before the crossing signal, I listen to the dog breathe. The sound is old and bronchitic, a dissonant issuing from this neat little body, the laboured wheezing of a young dog's breath. The man is fashionably dressed and the dog most probably loved and precious. I'm not sure if the dog is a French bulldog or a pug, but he's one of those who now form a widespread, snuffling, breathless band of canine respiratory distress. The lights change and man and dog walk off, the dog carrying his possibly malign genetic destiny, his future skin-fold pyoderma, the corneal ulceration which may affect his protruding eyes, the upper airway obstruction which is probably already causing him to wheeze. It's not the first time I've wondered – what made this man and others seek out and pay for creatures who they know may live shortened, suffering lives?

Deliberate selection for short limbs and long backs has caused dachshunds, shih-tzus, basset hounds and other breeds to suffer from the painful bone condition chondrodystrophy. Larger dogs such as Rottweilers, St Bernards and retrievers experience hip dysplasia, arthritis, osteosarcomas and degeneration of the joints. Eye problems are common in many breeds, as is deafness. Skin diseases and inflammation are caused by breeding for wrinkled skin in basset hounds, bloodhounds and Shar Peis. Blood, kidney, gastrointestinal and neurological ailments are common – many King Charles spaniels, griffons and chihuahuas suffer from the spinal-chord destroying syringomyelia, caused by having skulls too small to accommodate their brains. It is a condition that has increased greatly over the past twenty years and continues to do so. Cavalier King Charles spaniels also suffer from mitral valve disease while other heart conditions afflict boxers, Rottweilers and Dobermanns. Very small 'teacup' dogs suffer from increased bone fragility while 'flat faced' or brachycephalic dogs – pugs, bulldogs and Pekinese frequently suffer from brachycephalic obstructive airway syndrome (BOAS), which causes breathing difficulties and shortens their lives. Many dogs are artificially inseminated and as a result of selection for large heads and narrow pelvises are unable to give birth without a caesarean section.

Cats too suffer the results of breeding for 'desirable' traits, most often those associated with colour and appearance. Pedigree cats suffer disproportionately from dystocia – difficulty in giving birth, and subsequent high death rates for pedigree kittens. Manx cats may suffer from a number of ailments related to selection for short or no tails including spinal deformities, spina bifida and digestive problems. Scottish fold cats are subject to cartilage problems, leading to arthritic conditions while Burmese cats are prone to diabetes mellitus, cranial deformities, glaucoma and kidney stones. Both Burmese and Siamese cats may also suffer from BOAS, diabetes, asthma, lymphomas, strabismus, hip dysplasia and small intestinal adenocarcinomas. Rabbits such as the English 'lop' have significant health problems caused by their overlong ears. Selectively bred rats are subject to a number of health problems including greatly increased risk of tumours.

The small dog at the traffic lights is only one of many. Encouraged by online images, their popularity has increased to the point where, despite widespread publicity about their health problems, demand for them greatly exceeds supply, which has brought about not only the irresponsible breeding that produces unhealthy animals but has led to a huge increase in the hazardous and cruel 'farming' of dogs and their illegal trade and importation across borders. At least one danger of this trade is the possibility of the reintroduction of rabies as a result of faked certificates and importation of affected creatures. Images from puppy farms look remarkably similar to those from fur farms, showing the dirty, caged, abused and suffering creatures who we still continue to buy. What makes us do it? Why do we encourage a trade that exploits the sufferings of others? One suggestion is that the 'childlike' appearance of dogs such as pugs and bulldogs attracts us – according to a theory in evolutionary psychology, 'Kindchenschema', also known as 'neoteny', a positive response to the appeal of 'babylike' or 'cute' faces is an evolutionary way of ensuring the survival and nurturing of offspring. The theory may be correct (if you really think that bulldogs look like babies) but does not prevent us from making an ethical decision about who and what we buy. I watch at the traffic lights as the man leads the dog away, this lifelong victim of our desire for 'cute'.

No longer simply a matter of small, personal decisions, our animal owning has implications far wider than the privacy of our homes. It is increasingly subject to the moral, financial and political questions posed by knowledge of animal cognition and urgent considerations of consumption and resource. Feeding our pets involves similar questions to the ones we ask about feeding ourselves – what is healthy, affordable, necessary, ethical and environmentally sustainable? A study in 2017 assessed the environmental impact of companion animals in the United States. The findings were that dogs and cats were responsible for 25–30 per cent of the environmental impact of all meat consumption, that they created 64 million tons of carbon dioxide and methane and produced 5.1 million tons of faeces annually, the same as 90 million humans. The study suggested that, in the light of these figures, increasing pet keeping worldwide will make a huge, significant contribution to our current ecological crisis. (The suggestion that the food fed to animals is a 'by-product' of human food production is refuted by one study which points out that, increasingly, pets are being fed higher-quality meat and much of what is regarded as unfit for human consumption is considered so more on aesthetic than other grounds.)

As pet numbers increase, so do our purchases. Browsing through websites of pet products is like entering an anthropomorphized nightmare of overextended consumerism. One site offers 698 varieties of dog 'treats'. Another sells pet beer, wine and herbal tonics for pet anxiety. There are the luxurious beds, the electronic toys, the whimsical clothing. There are socks and shoes, hats, bow ties and dresses. There are shampoos, conditioners, dog-nail polish, fur dyes and whirlpool tubs. There are extensive ranges of veterinary psycho-pharmaceuticals to treat anxiety and behavioural problems, aromatherapy candles, colognes and 'fragranced' sprays to mask the creature's natural odours. There are the fancy-dress costumes, sharks or spiders, sumo wrestlers, light-up Halloween pumpkins and hundreds more.

Looking after the health of our pets may once have been simpler when treatments were limited and they had fewer complex problems. Now, in an endless cycle of concern and responsibility, we have to decide on the treatments and insurance, which may be too expensive for many pet owners, creating yet another division of privilege, an irreconcilable dilemma for those who cannot pay for treatments they know to be available for their beloved animals. Another decision is whether or not to have a newly acquired pet neutered. It may be a responsible action in limiting the future numbers of free-roaming animals such as cats, but while it may be convenient for owners, for the animals themselves there may be future health consequences for the animal in obesity, cancers or joint disease. We are embarrassed by the manifestations of our pet animal's sexuality, the subject usually being referred to through jokes or awkwardness, reflections of our reluctance to accept that however sensible the decision may seem, in terms of our own or their benefit, neutering is a denial of the natural right of another being. It is just another aspect of the total power we exercise over the lives of the animals we choose as companions. Writing in the poem 'Another Dog's Death' of the early spaying of his dog, John Updike describes her as knowing 'no nonhuman word for love'.

We expect so much from other species. For our purposes, they must be sufficiently like us for us to want to understand their behaviour and believe it very much like our own, but sufficiently unlike ourselves for us to be free of our concerns. They have to be easily sent to kennels when we wish to go on holiday, content to be left on their own all day, often confined in places much too small, or in conditions utterly unlike their natural habitats. In 1943, Elias Canetti wrote, 'It is not good that animals are so cheap.' He might have been writing about the hamsters, mice, rats, guinea pigs and gerbils frequently bought as suitable pets for children, some of whom will be loved, tended and eventually mourned, others of whom will be neglected and worse. Solitary creatures will be kept in pairs or groups, or social ones alone. Crepuscular or nocturnal creatures as many of them are will be expected to provide entertainment for diurnal children. Reluctantly, I remember school-gate conversations about unfortunate fates: the school rabbit forgotten over a summer when the parent who was to look after him went on holiday, the escaped mice, the hamsters who fell, disappeared, were drowned or squashed or found burned at the back of a gas fire. The incidents were invariably presented as amusing, told in a tone of mocking self-exculpation. I see a succession of online adverts selling unwanted hamsters and guinea pigs. The child for whom they were bought 'lost interest', the family is moving house, there was an accidental mating. ('Oops!') What they are being sold for, the cost of a cup of coffee, is the cost of another creature's life. What is the Umwelt of a puppy-farmed dog, a lone rat, desert gerbil, a Syrian hamster in a small plastic box?

What do we really know of the animals we buy? Our perceptions of their behaviour tell us that often they experience things in a similar way to ourselves and that we may describe their behaviour as love, anger, jealousy, delight, embarrassment, joy or grief because we have no other way to explain it. We all know what another creature's happiness or distress looks like because they look very much like our own. When we force explanations of their behaviour on them, 'She likes it!' when possibly she does not, or 'He doesn't mind' when clearly he does, we skew the relationship by manipulating an animal into being what we want. Other species possess 'intelligence' but too often we want it to be a mirror of our own. (Assessing intelligence in our own species is hard enough and the attempt to understand cognitive ability in other species is an unfinished and never-ending quest.) Even before I learned about the physical make-up of his remarkable brain and the research done into the high cognitive function of magpies, I knew that Spike, my magpie, was clever but I have never been able to explain satisfactorily how, without using terms that remove him from an avian world, which I will never fully comprehend, ones that would judge him by standards that were never his. I can say only that in observing everything he did and was, I knew him my equal and more likely not less than me but more.

Potential danger in other creatures is difficult to assess. We've all heard the dazed excuse 'I thought he wouldn't hurt a fly' expressed by the owner of the dog who kills or maims, the person who seems tragically unaware that a dog should not be expected not to hurt a fly or anything else, and that dog and victim should both have been prevented from either suffering or causing harm. In Jonathan Safran Foer's book *Eating Animals*, he writes of his relationship with his own dog and her 'foreignness', which includes being sufficiently unknown for him to feel uncertain that the dog wouldn't maul his baby. He is wary and sensible, unlike the advice I find on a website promoting the qualities of a particular breed of dog that suggests they are entirely suitable be left alone with children. Would anyone leave a creature of any sort, or indeed some humans, in a room alone with a small child? Ernest Hemingway would. In *A Moveable Feast*, a memoir of his life in Paris with his first wife, Hadley, he writes of her going out to play the piano while he – before going to sit in a cafe to drink and write – would settle their large cat in their son Bumby's cot. Despite warnings that it might be dangerous to leave the cat with the baby (objections Hemingway dismissed as 'ignorant and prejudiced') the baby survived the experience. Reassuring his readers that there was no need for a babysitter, Hemingway writes 'F. Puss was the baby-sitter.'

When we are considering the potential risk an animal may pose, appearance affects our judgement. Some dogs may be more belligerent than others, made so by training or treatment but for those who do not know a dog personally it may be difficult to tell. We bring our prejudices to the perception – some dogs are subject to the discrimination which afflicts us in our views of the *other* and often it is the owner who is judged. Staffordshire bull-terriers, 'Staffies', are particularly subject to negative views, often because of associations with cross-bred 'fighting' dogs. If people keep dogs who they believe have an air of menace it may be because they feel more protected in a particular dangerous world when they do. A favourite photo I have is of the one Staffie I knew well, a gentle and charming dog called Belle. In the photo she lies with an air of perfect happiness on the knee of her owner, my then very elderly grandfather-in-law. It is not a good photo – both man and dog have oddly glassy stares but the contentment and love it portrays transcends the technical shortcomings. Beauty seems a theme in the naming of these dogs – on Twitter, there appears the photo of another, similarly contented Staffie: 'Bella was particularly happy after rolling in goose poop,' the Tweet says, 'she is my heart.'

Considering the total dependency of domesticated and 'pet' animals on humans, the law professor and ethicist Gary Francione talks of the 'netherworld of vulnerability' to which they are subject. It is a vulnerability manifest in every facet of our dealings with them. The cruelties of every day spin out, major and minor, our national claims of 'love' often sounding hollowly over the cold ring of statistics, the 74,000 or so animals abandoned annually in Britain, the shameful list of prosecutions for hideous acts perpetrated daily against other species, the estimated 1.5 million abandoned 'shelter' animals killed annually in America, the 3,500 or so stray dogs killed in Britain. These are just the ones we know about. I'm driving one quiet Saturday afternoon in a street in a suburb of the city and see a woman walking with a dog stop, look around briefly before raising her foot and savagely kicking the dog's side.

In an essay, the American writer Alison Hawthorne Deming

remembers her cat, one of a feral litter found under the poetry centre where she works. The mother cat was fierce, disdainful of humans, 'like a war correspondent who has seen too much ever to believe in human kindness'. One of the staff tames the kittens and gives them away on the understanding that they'll be given literary names. Deming calls her cat after Christopher Smart's. A vocal lover of life, her cat lives as if 'each moment were his first on earth'. When one day the cat shows signs of neurological damage and tests positive for antifreeze, an agent commonly used for poisoning, Deming does not know if the act was deliberate or not, writing that she can imagine a neighbour doing it in irritation over some minor matter although she cannot be sure. The vet puts the cat to sleep and Deming reflects on the initial bitterness that encourages her to believe the mother cat wise in staying away from humans, a feeling she overcomes by remembering the happiness of the cat's life and her appreciation of his quality of innocent simplicity.

The vulnerability Francione writes about extends beyond the boundaries of species. Our love for others, human or not, makes us vulnerable, open to pain and loss, to the use and abuse of power and domination. Theodor Adorno's observation 'Love you will find only when you may show yourself weak without provoking strength' seems a disquieting reflection when thinking about some of our relationships with other species. Research into the link between violence towards animals and subsequent violence towards other people highlights the vicious chain which connects cruelty towards animals and children to domestic abuse. A relationship between children's cruelty to animals and later behaviour has long been recognized, anecdotal evidence at least suggesting that many of the world's most egregious serial killers and mass murderers were sadistic torturers of animals in childhood. John Locke's famous observation in 'Some Thoughts Concerning Education', in 1693, that children who 'ill use' animals may be prone later to do the same to people, appears to be substantiated by contemporary research. Studies which examine the connection between domestic violence and cruelty to animals are painful in their demonstration of how the bonds victims of domestic abuse may have with their pets are exploited as tools of control by violent partners: 'He said he would kill my dog if I left', 'he threw my kitten against the wall. He said he would do that to me and the children', 'He told the children he'd drown their rabbits', and worse. Where do the vulnerable go with their children and their pets? There are too few places of sanctuary, and fewer still that can accommodate animals. What are they to do?

In a remarkable piece published in the *New Yorker* in 2016, 'What Does a Parrot Know about PTSD?', Charles Siebert gives an account of his visit to Serenity Park, a parrot sanctuary set up in the grounds of a health centre for war veterans in California where scarred humans, wracked by post-traumatic stress disorder, care for abused and abandoned parrots. In this extraordinary partnership, humans and parrots provide mutual solace for the pain of their lives. Their life stories, for all their species difference, are similar in torment and memory: the birds' experience of abject cruelty at the hands of humans and the veterans, of the destructive, life-ruining aftermath of war. The parrots are, Siebert says 'twice traumatized' having been taken from their wild, natural lives to be subjected to captivity and cruelty. He writes of their flock behaviour, their vocal abilities, longevity and high-cognitive functioning, of the beautiful Goffin's cockatoo who was kept in a kitchen drawer and the Moluccan cockatoo abandoned in a mansion when the owners left. Others still call names and words in the languages of their previous owners and, like many of the veterans, display the agonizing behaviours of extreme distress as, in their slow progress towards healing, they transcend the imagined boundaries between species.

The relationships at Serenity Park are mutual, but in the wider field of 'assistance animals', human–animal interaction is more one-sided. A list I find which classifies the terms used for these animals sounds more like a catalogue of electrical equipment than one relating to living beings: *public-service animal, therapy animal, visitation animal, support animal*. What do we require of other species? That they work for us, aid us physically and emotionally, often while they are denied the right to any natural life? Should other species be expected to remedy and ameliorate the maladies of our bodies and souls? No doubt many human–animal relationships are of profound benefit to humans but call into question the kind of lives these animals will lead. Expectations are different from those demanded of pets, or 'companion' animals and many animals, bred for one particular purpose alone, may be destined to live in kennels all their lives.

These role differences create divisions. Some of the species we consider as 'companion' animals, although they may be the same species as ones not considered to be companions, have more protections than the others, but why? Establishing a hierarchy of rights gives licence to those who wish to treat other animals less well than we treat our pets. Should our grace alone confer benefit or protection? In his book *The Problem of Pain*, C. S. Lewis wonders if animals who have had a close relationship with a human might be granted immortality, 'not in themselves but in the immortality of their masters'.

Questions of immortality, or more correctly mortality, are always there. Years ago, I saw a photo in a magazine and although I haven't looked at it for a while, I still think about it. It is of a well-known modern American composer walking in Central Park in winter. In his arms he clutches a Labrador puppy who peers from inside his coat. It was an advert for a credit card or bank and although it may have been staged, man and dog both smile and seem happily at one. The aspect of time in the photo stays with me, the knowledge of our unequal lives, the lives of that man and that dog. We are overtaken by our beloved pets. Mary Oliver, that supreme writer about dogs, regrets their 'galloping' lives, harbouring an irrational sense that it is failure of our will and love that allows them to grow old. These loves measure our lives in steps and stages, in the ever-present reminder that we will be bereft, or they will be. Chicken is decades older than she would ever have been in the wild, but she is old and frail. One day in the local pet shop I fall into brief conversation with the owners whose pets are elderly chinchillas of eighteen and nineteen years old. We discuss the pleasures and sadness of taking care of old and much beloved pets. We accept the burden of our future grief and the decisions we might have to take as we are turned to gods in the powers of life and death.

In her essay 'The Fourth State of Matter' Jo Ann Beard writes about overlapping tragedies, of grief in sequence, the horrors of the murder of several of her colleagues – all space physicists – in a campus shooting at the University of Iowa, the failure of her marriage and her dread-laden anticipation of the death of one of her beloved dogs. She writes about her colleagues and their work of interpreting the universe with the gentle humour with which she describes her relationships and losses and her inability to face the question of ending the dog's life. Tending the dog with devotion through night-time anxieties and broken sleep, Beard is urged by some friends to end the dog's suffering while others reassure her that she will do it when she knows she must. After the shootings, it is the dog who provides comfort in a night of contemplating tragedy against the backdrop of the planets, Sirius the dog star, Jupiter's moon, the rings of Saturn – the area of study of one of her lost friends. 'My peer, my colleague . . .' Beard writes of the dog, 'we used to call her the face of love.'

My magpie died and parrots and rats and doves died. The garden is full of small burial sites, places where a creature who has been part of, and contributor to our lives, has been placed, some remembered clearly, some less so. These ceremonies and memorials are what we may or may not do in the acknowledgement of their lives and deaths. Some are marked and some not. There is a tiny granite slab, carved by a friend with a gilded 'C' to mark the grave of a rat, but some have plants instead, a deep red *Heuchera*, a white *Potentilla*, a rose. For the magpie, there's a large beach stone of fine, smooth grey.

I am not around one afternoon when a dove is killed by a hawk.

Towards evening, I notice a spread of white on the grass, the hawk still in the middle of the feathered circle, plucking. As I'm watching, the young magpie whose progress I have followed all spring flies down and begins to circle the diner's periphery, hovering, peering. In time, the hawk lifts from the grass, flies low over the garden wall, disappears. Only then, the magpie begins to run around the circle of feathers and bones, calling, continuing for a few minutes before he too flies off. Is this the behaviour of distress? Is it curiosity or caution? We may not be the only ones who lose and grieve and mourn. Magpies, among other creatures, mark death by ceremony, although we cannot know what the ceremony means. They have been seen to tear grass, which they lay over the corpse of a fellow magpie. Whales and dolphins appear to display grief, particularly in the loss of young. An orca was observed carrying a dead calf for seventeen days in the Salish Sea. Giraffes, primates, birds and elephants have all been seen to display various kinds of 'mourning' behaviour, staying close to or maintaining vigil, cleaning the corpses, trying to resuscitate the dead and while it is difficult to say with certainty what any other living being thinks or feels, their behaviour is sufficiently like ours to be able to consider that it might be grief. I have seen what looks like grief in birds. I have seen doves who displayed apparent grief after the death of a mate crouching into themselves, shivering, apparently bereft.

The hawk did not come back to finish the meal. The next day, I collected what was left and dug a small hole in the earth and as I put the dove-bits in wondered what it is about a handful of wet feathers, a scrape of flesh and bone which provokes the need to do it. Might we be extending ourselves as far as to hope that someone might do the same for us? For this one, there is no ceremony, only a small placing, the commonplace marking of a finished life.

In spring, the family come for Passover. Although we are not strict observers of the very many tenets of our religion, we observe this festival, and specially the Seder meal, with dedication. We believe it to be important, and increasingly so. It is timely and all too urgent in its reminders and in the lessons it should teach us about freedom and oppression and the treatment of others. We read the ancient texts of the Haggadah as we always do. Over the past few years, we have incorporated new additions, specially written for the festival so that we can reflect on connections between past and present. This year, we include one from a human rights organization and one from Marion Nestle, an American professor of Nutrition, Food Studies and Public Health. Both discuss social justice, ethics and freedom from oppression. Professor Nestle writes about hunger and food equality, about employment rights, animal welfare and sustainability. The topics lead on to discussion about food and the rights of other species and all the other current, worryingly persistent topics that are occupying everybody's mind.

Over the few days we spend together, as we sit around drinking coffee and wine and talking, a small bird is with us. She may be on the top of a door or hanging upside down from a window blind, sitting on a knee or shoulder. She may well be dancing. The bird, whose name is Pocket, is the size of a sparrow. A green-cheeked conure, *Pyrrhura molinae*, she is a bird with feathers the colours of the flowers and foliage of a soft, pink moss rose, owner of the most formidable character and passions that I have ever encountered in a member of any species, human or not. She is companion bird to Bec and Leah, successor to the formidable cockatiel who, from her childhood, for twenty-two years, was the companion of Bec's life.

In his book *Parrot Culture: Our 2500-Year-Long Fascination with the World's Most Talkative Bird*, Bruce Thomas Boehrer presents both a wonderfully comprehensive cultural history of parrots and, as a dedicated parrot-lover himself, an affectionate and heartfelt portrayal of parrots and their present plight. He writes of the beauty and self-evident cleverness that made these birds attractive to humans in the first place, traded, exchanged, valued, prized throughout the ancient world, creatures of fascination until today. In describing what he knows about parrots personally, he uses vocabulary that anyone worried about accusations of anthropomorphism might hesitate to use but in describing parrots there are no other words. 'Headstrong' is one, 'defiant, jealous, remorseless, implacable, bloody-minded, emotionally needy' are others. In suggesting that parrots expect to be treated as equals, he is entirely correct. With highly developed brains and consequent high levels of cognition, they are capable of levels of behavioural sophistication that are disquieting when compared with most common views of other species. Boehrer writes about the many unmet emotional demands and expectations that frequently lead to parrots being badly treated and neglected. They are, he suggests, 'just too intelligent for their human owners'.

Knowing that this is so, every effort is made to ensure that Pocket's demands are catered to and her wishes respected. Independent-minded and certain, she is part of every conversation. She is assiduous in saying 'Thank you' when given a dish of raspberries or pomegranate seeds. She flies through the kitchen to visit Ziki the crow with cries of 'Hello, baby!' but Ziki, alas, is unenthusiastic about her attentions and takes refuge by the back door. Once, she would groom Chicken's neck feathers when allowed, but now she understands that Chicken is old and weary and leaves her alone. When she bathes in the big dish of water we give her, we remember that she is actually a theropod dinosaur. She huddles on a lampshade to try to dry herself. (All birds wear their dinosaurian origins close under the cover of their feathers. They are smaller when they're wet, the structure and thinness of their necks more obvious. Then, they are just tiny, wet dinosaurs trying to get warm.) Her favourite dancing music is the Darth Vader theme from *Star Wars* and 'Nights in White Satin' by the Moody Blues and her method of dancing, more march than dance, demonstrates the cool brilliance that only parrots seem to possess.

This bird was bought from a breeder in Scotland to fulfil my granddaughter Leah's wish to have a parrot, as her mother did. When she was deemed to be old enough and responsible enough to participate in caring for one, the parrot was bought. We are respectful of her origins because, as Bruce Thomas Boehrer point out, parrots are only a few generations away from 'their 35 to 38 million years of wild inheritance'.

The illegal trade in parrots is destroying parrot populations worldwide. One study demonstrates that since 1992 the population of grey parrots in Ghana has fallen by 90–99 per cent as a result of both the illegal pet trade and habitat loss. Millions of birds belonging to at least 259 species of parrots are known to have been traded in the past few decades while some 30 per cent of the almost four hundred known psittacine species are threatened with extinction. The criminal trade in parrots is particularly cruel, with huge numbers of birds dying during capture and transportation. By now, many parrot species are already extinct. A recent study suggests, unsurprisingly, that the parrots most at risk from the illegal trade are 'the most attractive species'.

While we are sitting around discussing parrots and the nature of the world, it is still spring, but later in the summer, a fire at Ometepe in Nicaragua, started deliberately or by carelessness, will destroy a wildlife refuge, home to fifty species of birds, including the rare, endangered yellow-naped Amazon parrot, already prey to wildlife poachers. Other creatures will perish, armadillos, howler and capuchin monkeys, agoutis, sloths, bats and reptiles. Fires will ravage Brazil. They will destroy habitats, forests, worlds. The habitat of Pocket's family, her ancestors, her culture will be destroyed in the firestorms caused by human greed. Millions of animals, birds and insects will die and ecosystems will be destroyed. Of the 1,500 or so bird species who live in rainforest environments, many will be affected if not killed. The scarlet and hyacinthine macaws, the hoatzins, the jenday conures, the toucans and king vultures, the golden parakeets, Belem curassows, the black-winged trumpeters, the veeries, the blackpoll warblers will be consumed in the flames with which our species continues to destroy the earth.

But here, in spring, the garden stirs in temperate ease. As we talk, small birds crowd the bird table. On the roof of the room we still call the rat-room in tribute to the pet rats whose domicile it was long ago, a handsome wood pigeon sedulously grooms her feet. From below the window, a magpie, earnest and engrossed, mutters words to herself whose meaning only she can know. In the study, Chicken moves with the sunlight, following its trajectory around the floor, in this what I fear may be her last spring.

And now we extend our discussion to the delights and the demerits of pet-keeping. The issues involved have become more complex in the years since we began with two white doves, a cockatiel and a rook and none of us can be sure that we would do again what we have done in keeping creatures. Although we try to reassure ourselves that we have done the best we can for them, and that they have always appeared happy, however a human can judge, the entire rationale of our thinking has to suggest, these many years on, that that might not be good enough. We've borrowed their lives and the debt seems one beyond our paying back. We talk about what we could never have known without them, about the empathy of creatures, their learning, their fierce intelligence, their memory, affection, vulnerability, their unswerving faithfulness to who they are.

After everyone goes home, I go through the notes and books and assortment of references I've gathered up while I have been working over the past few years. There is Vasari's writing about Leonardo da Vinci: 'often when he was walking past the place where birds were sold he would pay the price asked, take them from their cages and let them fly off into the air, giving them back their lost freedom'. There is Loren Eiseley describing encounters with wild creatures: the muskrat, 'an edge of the world dweller' who seems too trusting for his own good, appearing at Eiseley's feet 'with his little breakfast of greens' and the dance he does with a crane in Philadelphia Zoo, the young fox he plays with, tumbling 'round and round for one ecstatic second', what he describes as 'the gravest, most meaningful act I shall ever accomplish'. There is a photo I found of Claude Lévi-Strauss with a crow on his shoulder. He, like Elias Canetti, wished that he could form a close relationship with another species. In them, the line dissolves. While living in Connemara in 1948, Ludwig Wittgenstein developed an interest in and love for birds. In her elegiac memoir of a beloved friend in New York, Sarah Manguso describes the article she has to write about services of blessing for animals in Manhattan in the days after 9/11 and of: 'all the depressed rescue dogs' being taken to church and led to the altar, of a bald eagle leading one procession followed by police and rescue dogs, a bloodhound named Chase waiting at the altar to receive his blessing. I find the epitaph to a horse on a Roman grave monument: 'you, who outran the wandering birds and beat the winds, graze no longer in the Tuscan or Siculan woods but in this tomb . . .'

It is late afternoon and my cousin Rog and I are strolling on a rocky beach on the day after the summer solstice, the moment of time's turning on itself, the beginning of the dwindling of the light. At this hour, it melts from crystal to pearl in a fine shimmer over an easy summer sea. We walk like this every year when Rog visits, exploring the steep stone harbour villages, the small, elegant Georgian towns along the coast. As we follow the curve of the headland, there is a low patch of stones and fluttering pennants to our right and for a moment, we can't identify what it is. Approaching, we see that it is a small graveyard. Death in miniature, crowded with tiny plots, it is a graveyard for pets. We make our way carefully around it for a while, looking at the pictures, toys and painted stones, the miniature carved granite headstones, the flags and flowers. We read the names and quiet expressions of loss in this small, windswept monument to love.

When we get home, Chicken calls from the study. I go in to greet her. The evening light shines behind the imprint of the dove's wings on the glass. 'The Angel of History' seems indelible, an image of flown doves and coming storms and I know it will not be washed away, even by the autumn rain.

Acknowledgements

Many people have given me generous help during the writing of this book. One or two have asked me not to use their names publicly and so I have thanked everybody personally and privately.

I am as ever, deeply grateful to my wonderful agent Jenny Brown whose support, wisdom and enthusiasm I continue to appreciate beyond measure.

My special thanks to Max Porter for believing this book to be a good idea from the start. I thank him too for his help, advice and encouragement and wish him high flight for his undoubtedly stellar career in the future. Many, many thanks to my editor Laura Barber for her thoughtful expertise and patience, to Pru Rowlandson and everyone at Granta with whom it is a continuing pleasure and privilege to work.

I am grateful to my beloved family, which includes of course, Pocket, the green-cheeked conure and Ziki the crow who, among others, inspired me to see beyond the boundaries. My greatest debt will always be to Chicken the rook who was beside me during the entire writing of the book.

Great gratitude goes too to Aberdeen Central Library whose staff have, with cheerful grace, helped me in every way they can, descending more times than is reasonable to the 'Reserve stock' section in the basement to unearth the obscure and ancient books for which I've asked them. Both library and staff represent the absolute best of these most precious, valuable and necessary of institutions.

Bibliography

1. SHARING A PLANET

David B. Williams, 'Benchmarks: September 30 1861: Archaeopteryx is discovered and described', *Earth Magazine*, September 2011.

Kundrát, Nudds et al., 'The first specimen of Archaeopteryx from the Upper Jurassic Mörnsheim Formation of Germany', *Historic Biology*, 31 (1), 24 October 2018.

Cynthia Marshall Faux and Kevin Padian, 'The Opisthotonic Posture of Vertebrate Skeletons: Postmortem Contraction or Death Throes?', *Paleobiology*, 33 (2), Spring 2007.

Lida Xing et al., 'A flattened enantiornithine in mid-Cretaceous Burmese amber: morphology and preservation', *Science Direct*, 63(4), 28 February 2018.

Lida Xing et al., 'A mid-Cretaceous enantiornithine (Aves) hatchling preserved in Burmese amber with unusual plumage', *Gondwana Research*, 49, September 2017.

Marcia Bjornerud, *Timefulness: How Thinking Like a Geologist Can Help Save the World* (Princeton University Press, New Jersey, 2018).

Bruno Latour, *We Have Never Been Modern*, trans. Catherine Porter (Harvard University Press, Cambridge, Mass., 1993).

Bruno Latour, 'Agency at the Time of the Anthropocene', *New Literary History*, Vol. 45, 2014.

Alaina Levine, *The Large Horn Antenna and the Discovery of Cosmic Background Radiation* (American Physics Society, 2009).

Erik M. Leitch, 'What is the Cosmic Background Radiation?', *Scientific American*, 2004.

Leah Poffenberger, '2019 Nobel Prize in Physics', *American Physics Society News*, October 2019.

Elizabeth Howell, 'What is the Big Bang Theory?', Space.com, 7 November 2017.

Zaremba Niedzwiedzka, Caceres et al., 'Asgard archaea illuminate the origin of eukaryotic cellular complexity', *Nature* 541, 11 January 2017.

Joshua A. Krisch, 'Asgard Archaea Hint at Eukaryotic Origins: A newly discovered superphylum of archaea may be related to a microbe that engulfed a bacterium to give rise to complex eukaryotic life', *The Scientist*, 18 January 2017.

Anja Spang et al., Complex archaea that bridge the gap between prokaryotes and eukaryotes, *Nature* 521, 6 May 2015.

Paul Rincon, 'Newly found microbe is close relative of complex life', BBC News/science-environment, May 2015.

Madeline C. Weiss, 'The last universal common ancestor between ancient Earth chemistry and the onset of genetics', *PLoS Genetics*, 16 August 2018

Keith Cooper, 'Looking for Luca', Astrobiology at NASA, March 2007.

'Great Oxidation Event: More oxygen through multicellularity', University of Zurich, *ScienceDaily*, 17 January 2013.

Stephen Jay Gould, *Wonderful Life: Burgess Shale and the Nature of History* (Vintage, New York, 2000).

Russell J. Garwood et al., Royal Society Publishing, 2016. The fossil record and taphonomy of butterflies and moths, in 'The fossil record and taphonomy of butterflies and moths (Insecta, lepidoptera): Implication for evolutionary diversity and divergence-time estimates', Jae-Cheon Sohn et al., BMC, *Evolutionary Biology*, 15 (12), 4 February, 2015.

Jae-Cheon Sohn et al., 'The fossil record and taphonomy of butterflies and moths (Insecta, Lepidoptera): implications for evolutionary diversity and divergence-time estimates', BMC, Evolutionary Biology 15 (12), 2015.

Conrad C. Labandeira et al., 'The evolutionary convergence of mid-Mesozoic lacewings and Cenozoic butterflies', Proceedings of the Royal Society B, 10 February 2016.

Mike Gray, *Spider Origins* (Australian Museum, 2018).

Russell J. Garwood et al., 'Almost a Spider: A 350-million-year-old fossil arachnid and spider origins', *Royal Society Publishing*, 30 March 2016.

Michael Marshall, 'The Evolution of Life', *New Scientist*, 2009.

Stephen L. Brusatte, Jingmai K. O'Connor, Erich D. Jarvis, 'The Origin and Diversification of Birds', *Current Biology*, 25 (19), 5 October 2015.

Stephen L. Brusatte, 'Evolution: How Some Birds Survived When All Other Dinosaurs Died', *Current Biology Dispatches*, 26 (10), 23 March 2016.

Colin Barras, 'Ancient Humans: What We Know and Still Don't Know About Them', *New Scientist*, May 2017.

John Pickrell, 'Timeline of Human Evolution', *New Scientist*, 2006.

Jean-Jacques Hublin, Abdelouahed Ben-Ncer, 'New fossils from Jebel Irhoud, Morocco and the pan-African origin of Homo sapiens', *Nature* 546, 2017.

Matthew Warren, 'Biggest Denisovan fossil yet spills ancient human's secrets', *Nature* 569 (16–17), 1 May 2019.

Ian Sample, 'Piece of Skull Found in Greece is "oldest human fossil outside Africa"' – Remains discovered on Mani peninsula could rewrite history of Homo Sapiens in Eurasia, *Guardian*, 10 July 2019.

Chris Stringer, 'The New Human Story – Master of Science/Where We Come From', *Financial Times* Weekend Magazine, July 2019.

Cullerlie Stone Circle, Garlogie, Aberdeenshire; www.historicenvironment.scot

'A New Light on the Earliest Neolithic in the Dee Valley, Aberdeenshire', *PAST*, the Newsletter of the Prehistoric Society, No. 50, July 2005.

British Geological Survey timeline of geology; bgs.ac.uk.

David Leveson, *A Sense of the Earth* (The Natural History Press, New York, 1971).

Lauret E. Savoy, Eldridge M. Moores and Judith E. Moores (eds.), *Bedrock: Writers on the Wonders of Geology* (Trinity University Press, Texas, 2006).

Jacquetta Hawkes, *A Land* (Collins Nature Library, 2012).

James Hutton, *Theory of the Earth* (1795). *Theory of the Earth* was published first as two papers in 1788 and later, in expanded form, in two volumes in 1795.

The quotation is from *The Bride of Messina* by Johann Christoph Friedrich von Schiller (1759–1805).

Carly Hilts, 'Mesolithic Timelords: A monumental hunter-gatherer "clock" at Warren Field, Scotland', *Current Archaeology*, 5 September 2013.

Indrani Mukherji, Ross R. Large et al., 'The Boring Billion, a Slingshot for Complex Life on Earth', *Scientific Reports*, 8 (4432), 13 March 2018.

University of California Museum of Paleontology Geologic Time-scale; ucmp.berkeley.edu

'Timeline of Earth's Average Temperatures Since the Last Ice Age Glaciation', Bulletin of the Atomic Scientists; the bulletin.org

David Balto, 'What Thawed the Last Ice Age?', *Scientific American*, April 2012.

Michael Marshall, 'The History of Ice on Earth', *New Scientist*, May 2010.

Michael Slezak, 'Key Moments in Human Evolution Were Shaped by Changing Climate', *New Scientist*, September 2015.

For information about changes in CO_2: Records of Mauna Loa Observatory, Hawaii, and 350.org

Jacob L. Weisdorf, 'From Foraging to Farming: Explaining the Neolithic Revolution', *Journal of Economic Surveys*, 19 (4), February 2005.

Jacob L.Weisdorf, 'Why did the first farmers toil? Human metabolism and the origins of agriculture', *European Review of Economic History*, 13 (2), August 2009.

'The Mesolithic-Neolithic transition in Scotland: ways forward' – discussion paper, Scottish Archaeological Research Framework.

Ofer Bar-Yosef and Anna Belfer-Cohen, 'The Origins of Sedentism and Farming Communities in the Levant', *Journal of World Prehistory* 3, (4) December 1989.

Edwin Morgan's beautiful poem 'The Archaeopteryx's Song' is published in *Edwin Morgan: Collected Poems* (Carcanet, 1996).

Zhen-Tao X., Stephen, F. R., & Yao-Tiao, J., 'Astronomy on Oracle Bone Inscriptions', *Quarterly Journal of the Royal Astronomical Society*, 36, 1995.

Walter Benjamin, 'Theses on the Philosophy of History', 1940.

2. A THING APART

Perrice Nkombwe XXIII, 'Prehistoric Art and Museology: the Case of the Livingstone Museum, Zambia', Valcomonica Symposium, 2009.

H. S. Groucutt, et al., Rethinking the dispersal of Homo sapiens out of Africa, *Evolutionary Anthropology*, 24 (4), July/August 2015.

Huw S. Groucutt, Rainer Grün et al., 'Homo sapiens in Arabia by 85,000 years ago', *Nature Ecology and Evolution*, 2, 9 April 2018.

Axel Timmerman and Tobias Friedrich, 'Late Pleistocene climate drivers of early human migration', Nature, 538 (92–5), 21 September 2016.

Paul S. Martin, *Pleistocene Extinctions: The Search for a Cause* (Yale University Press, Conn., 1967).

Michael Balter, 'What Killed the Great Beasts of North America?', *Science* (The American Association for the Advancement of Science, 2014).

David Melzer, 'Pleistocene Overkill and the North American Mammalian Extinctions', *Annual Review of Anthropology*, 44 (33–53), October 2015.

Robert Sussman and Joshua Marshack, 'Are Humans Inherently Killers?', Global Nonkilling Working Papers, Center for Global Nonkilling, 2010.

S. A. Elias, D. Schreve, 'Late Pleistocene Megafaunal Extinctions: Can We Find the Smoking Gun?', in Scott Elias (ed.), *Encyclopedia of Quaternary Science*, 2nd edition, Elsevier, 2013.

Christopher Johnson, 'Hunting or Climate Change? Megafauna Extinction Debate Narrows', The Conversation, 2012.

J. Victor Moreno-Mavar, Ben A. Potter, et al., 'Terminal Pleistocene Alaskan genome reveals first founding population of Native Americans', *Nature*, 553 (203–7), 3 January 2018.

Todd Surovell, 'Extinctions of big game', *Encyclopaedia of Archaeology* (Elsevier, 2018).

Gary Haynes, 'A review of some attacks on the overkill hypothesis, with special attention to misrepresentations and doubletalk', *Quaternary International*, 169 (84–94), 2 July 2007.

Mathias Stiler, Gennady Baryshnikovet et al., 'Withering Away – 25,000 years of Genetic Decline Preceded Cave Bear Extinction', *Molecular Biology and Evolution*, 27 (5), May 2010.

Felise A. Smith, Rosemary E. Elliott Smith et al., 'Body Size downgrading of mammals over the late Quaternary', *Science*, 360 (6386), 20 April 2018.

Lisa Nagaoka, Torben Rick and Steve Wolverton, 'The overkill model and its impact on environmental research', *Ecology and Evolution*, 8 (19) October 2018.

Lisa Nagaoka, 'The Overkill Hypothesis and Conservation Biology', in *Conservation Biology and Applied Zooarchaeology*, Steve Wolverton and R. Lee Lyman (eds.) (University of Arizona Press, Tucson, AZ., 2012).

Surovell, Todd, et al., 'Test of Martin's overkill hypothesis using radiocarbon dates on extinct megafauna', Proceedings of the National Academy of Sciences of the United States of America, 2016.

Judith Thurman, 'First Impressions: What Does the World's Earliest Art Say About Us?', *New Yorker*, June 2008.

Jeff Tollefson, 'Human evolution: Cultural roots – A South African archaeologist digs into his own past to seek connections between climate change and human development', *Nature*, 4982 (7385), 15 February 2012.

Chip Walter, 'Origins of Art: First Artists', *National Geographic*, January 2015.

Werner Herzog, director of *The Cave of Forgotten Dreams*, a 2010 film.

John Berger, director of *Dans le Silence de la Grotte Chauvet*, a 2002 film for Arte France.

John Berger, 'Past present', article on the Chauvet caves, *Guardian*, 12 October 2002.

'Hand Stencils in Upper Paleolithic Cave Art', A research project of the Department of Archaeology, Durham University.

Vulture bone and mammoth ivory flutes, possibly 40,000 years old, were found in 2008 at Hohe Fels, a Stone Age cave in Southern Germany, by a team led by Nicholas Conrad of the University of Tübingen.

C. S. Henshilwood, F. d'Errico, K. L. van Niekerk et al., 'Oldest pigment factory dates back 100,000 years', *Science*, 2011.

Günter Berghaus (ed.), *New Perspectives on Prehistoric Art* (Praeger, Conn., 2004).

Cennino d'Andrea Cennini, trans. Daniel V. Thompson Jr., *The Craftsman's Handbook – Il Libro dell'Arte* (Dover, New York, 1960).

Susan Brind Morrow, *The Names of Things: A Passage in the Egyptian Desert* (Riverhead Books, New York, 1997).

Elon Gilad, *Genesis of Genesis, Haaretz*, October 2015.

Barry B. Powell, 'The Big Bang is Hard Science. It is also a Creation Story', Nautilus, 8 August 2019.

Bronislaw Malinowski, *Myth in Primitive Psychology* (Doubleday, New York, 1955).

Catherine Feher Elston, *Ravensong – A Natural and Fabulous History of Ravens and Crows* (Northland Publishing, Arizona, 1991).

Bill Reid and Robert Bringhurst, *The Raven Steals the Light* (University of Washington Press, Seattle, 1996).

Kerry McCluskey, *Tulugaq: An Oral History of Ravens* (Inhabit Media, Toronto, 2013).

Lisa Maher et al., 'A Unique Human-Fox Burial from a Pre-Natufian Cemetery in the Levant (Jordan)', PLoS ONE6 (1), 26 January 2011.

Simon J. M. Davis, François R. Valla, 'Evidence for domestication of the dog 12,000 years ago in the Natufian of Israel', *Nature*, 276 (608–10), 7 December 1978.

Grünberg J. M., 'Animals in Mesolithic Burials in Europe', *Anthropozoological*, 48 (2), 13 September 2013.

Luc Janssens et al., 'A New Look at an Old Dog: Bonn-Oberkassel Reconsidered', *Journal of Archaeological Science*, 92, April 2018.

P. B. Pettit et al., 'The Gravettian burial known as the Prince ("Il Principe"): new evidence for his age and diet', *Antiquity*, Cambridge University Press, 2015.

Formicola V., Milanesi Q., Scatsinis, C., 'Evidence of Spinal Tuberculosis at the beginning of the fourth millennium BC from

Arene Candide cave (Liguria, Italy)', *American Journal of Physical Anthropology*, 72 (1), 1987.

Julien Riel-Salvatore, Claudine Gravel-Miguel et al., 'New Insights into the Paleolithic Chronology and Funerary Ritual of Caverna delle Arene Candide', in Valentina Borgia, Emanuela Cristiani (eds.), *Palaeolithic Italy: Advanced studies on early human adaptations in the Apennine peninsula* (Sidestone Press, 2018).

Erica Hill, 'Archaeology and Animal Persons: Toward a Prehistory of Human-Animal Relations, Environment & Society', *Advances in Research* (Berghahn Books, Oxford, 2013).

Anne-Sofie Gräslund, 'Dogs in Graves – a Question of Symbolism?', Pecus. Man and Animal in Antiquity: Proceedings of the Conference at the Swedish Institute in Rome, September 9–12, 2002, 2004.

Melinda Zeder, 'Domestication and early agriculture in the Mediterranean Basin: Origins, diffusion, and impact', Proceedings of the National Academy of Sciences, 105 (33), 2008.

Overton N. and Hamilakis, Y., 'A Manifesto for a social zooarchaeology: swans and other beings in the Mesolithic', *Archaeological Dialogues*, 20 (2), 2013.

Kristiina Mannermaa, 'On whooper swans, social zooarchaeology and traditional zooarchaeology's weight', published online by Cambridge University Press, 8 November 2013. https://doi.org/10.1017/S1380203813000196

Christopher Moreman, 'On the relationship between birds and spirits of the dead', *Society and Animals*, 22 (5), 2014.

Brenda Martin, Ph.D., Kate Bowell, Treloar Tredennick Bower, and Terry Burton, 'Excavation of Lindenmeier: A Folsom Site Uncovered 1934–1940'. Additional research by Lesley Drayton. Fort Collins Museum & Discovery Science Center, 2009.

Loren Eiseley, *Notes of an Alchemist* (Charles Scribner's Sons, New York, 1972).

3. SOULS

Michael Frammartino directed *Le Quattro Volte* (New Wave Films). The film won the Best European Film award, and the starring actor, the dog Vuk, won the Palme Dog award at the 2010 Cannes Film Festival.

The Soul in Islamic Philosophy, www.muslim philosphy.com

Understanding the Three Types of Nafs, Zaynab Academy, zaynabacademy.org

Jelaluddin Rumi, *Look! This is Love: Poems of Rumi*, translated by Annemarie Schimmel (Shambhala Centaur Editions, Boston, Mass., 1996).

Judah Halevi, 'To the Soul', trans. T. Carmi (ed.), *The Penguin Book of Hebrew Verse* (Allen Lane, London, 1981).

Catholic Answers website: 'Animals and plants can't do anything which transcends the limitations of matter. Although some animals seem clever, they don't actually possess conceptional intelligence. They can't, for instance, conceive of the abstract notion of justice . . .'

Bertrand Russell, *A History of Western Philosophy* (Unwin Paperbacks, London, 1984).

D'Arcy Wentworth Thompson, *On Growth and Form* (Cambridge University Press, 1961).

D'Arcy Wentworth Thompson, *On Aristotle as a Biologist*, The Herbert Spencer Lecture, University of Oxford, 14 February 1913.

Aristotle, *The History of Animals*, trans. D'Arcy Wentworth Thompson (privately published, 2019).

Aristotle, *De Anima, On the Soul*, trans. Hugh Lawson-Tancred (Penguin Classics, London, 1987).

Diogenes Laertius, *Lives of Eminent Philosophers*, R. D. Hicks (ed.), 1972; perseus.tufts.edu.

Moses Hadas (trans.), *The Stoic Philosophy of Seneca, Essays and Letters* (Norton, New York, 1990).

Alexander Philonis, Abraham Terian, *De Animalibus* (Scholars Press, Chicago, 1981).

Celsus, *On the True Doctrine: A Discourse Against the Christians*, trans. R. Joseph Hoffman (Oxford University Press, USA, 1995).

Thomas F. Bertonneau, 'Celsus, the First Nietzsche: Resentment and the Case Against Christianity', *Anthropoetics* III, No.1, Spring/Summer 1997.

Plutarch, *De sollertia animalium*, as published in Vol. XII of the Loeb
　　Classical Library edition, 1957.
Porphyry, *On Abstinence from Animal Food*, Book I.
Jelbert, S. A., Taylor, A. H., Cheke, L. G., Clayton, N. S., Gray, R. D.,
　　'Using the Aesop's fable paradigm to investigate causal understanding of
　　water displacement by New Caledonian crows', PLOS One 9 (3), 2014.
Daniel Hillel, *Out of the Earth: Civilization and the Life of the Soil*
　　(University of California Press, Berkeley, 1992).
Genesis – *The Jewish Encyclopedia*, Vol. 5 (Funk and Wagnalls, New
　　York, 1925).
Moses Maimonides, *Guide for the Perplexed* (Dover Press revised edition,
　　2000).
Reza Gharebaghi, Mohammad Reza Vaez Mahdavi et al., 'Animal
　　Rights in Islam', AATEX, 14 August 2007.
Summa Theologica of St Thomas Aquinas, 5 vols. (Christian Classics,
　　English Dominican Province Translation Edition, 1981).
St Augustine, *The Catholic and Manichaean Ways of Life* (The Fathers
　　of the Church, Volume 56), trans. Donald A. Gallagher, Idella J.
　　Gallagher (CUA Press, Washington DC, 2010).
Lynn White, Jr., 'The Historical Roots of Our Ecological Crisis', *Science*,
　　New Series, Vol. 155, No. 3767, pp. 1203–7, American Association
　　for the Advancement of Science, 1967.
Michael Paul Nelson, Thomas J. Sauer, 'The Long Reach of Lynn White
　　Jr's, "The Historical Roots of Our Ecologic Crisis"', *Nature, Ecology
　　and Evolution*, March 2016.
Keith Thomas, *Man and the Natural World: Changing Attitudes in
　　England 1500–1800* (Penguin, London, 1984).
Michel de Montaigne, *The Complete Essays* (Penguin, London, 2003).
René Descartes, *Discourse on Method, and the Meditations*, F. E. Sutcliffe
　　(ed.) (Penguin, London, 1968).
René Descartes, *Passions of the Soul and Late Philosophical Writings*,
　　trans. Michael Moriarty (Oxford University Press, 2015).
Jonathan Bennett, 'Correspondence between Descartes and Princess
　　Elisabeth of Bohemia', 2017; earlymoderntexts.com
John Cottingham, 'A brute to the brutes? Descartes' treatment of
　　animals', *Philosophy*, 53 (206), 551–9, 1978.

Tim Ingold, *Lines: A Brief History* (Routledge, London, 2007).

Val Plumwood, 'Human Exceptionalism and the Limitations of Animals: A Review of Raymond Gaita's "The Philosopher's Dog"', *Australian Humanities Review*, 2007.

Mary Midgley, *On Being an Anthrozoon* (Centre for Humans and Nature, 2012).

Mary Midgley, *On Not Needing to be Omnip*otent (Centre for Humans and Nature, 2012).

Frans de Waal, *Are We Smart Enough to Know How Smart Animals Are?* (Granta, London, 2016).

A Singular Species: 'The Science of Being Human', *Scientific American*, September 2018.

Website devoted to the promotion of the idea of human exceptionalism: Discovery Institute–www.discovery.org

Joyce E. Salisbury, 'Do Animals Go to Heaven? Medieval Philosophers Contemplate Heavenly Human Exceptionalism', *Athens Journal of the Humanities and Arts*, January 2014.

Sy Montgomery, *The Soul of an Octopus: A Surprising Exploration into the Wonder of Consciousnes*s (Simon and Schuster, New York, 2015).

4. BLOOD

Jewish Museum, London, the exhibition 'Blood, Uniting and Dividing', 5 November 2015–28 February 2016.

Sacrifice – *Jewish Encylopedia*, Vol. 10 (Funk and Wagnalls, New York, 1925).

Henri Hubert and Marcel Mauss, *Sacrifice: Its Nature and Functions* (University of Chicago, Chicago, 1981).

Emily Wilson, *Seneca: A Life* (Allen Lane, London, 2015).

Metamorphosis, Book XV, Ovid.

Plutarch, *De sollertia animalium*, as published in Vol. XII of the Loeb Classical Library edition, 1957.

Porphyry, *On Abstinence from Animal Food*, Book I.

Rob Dunn, 'Human Ancestors Were Nearly All Vegetarians', *Scientific American*, 23 July 2012.

Jo Day, 'Botany meets archaeology: people and plants in the past', *Journal of Experimental Biology*, 64 (18), December 2013.

Briana Pobiner, 'Evidence for Meat-Eating by Early Humans', Human Origins Program, Smithsonian Institution, Nature Education, 2013.

Yoel Melamed, Mordechai Kislev, et al., 'The Plant Component of an Acheulian diet at Gesher Benot Ya'aqov, Israel', PNAS 113 (51), 5 December 2013.

Peter Ungar, 'The "True" Human Diet – From the standpoint of paleoecology, the so-called Paleo diet is a myth', *Scientific American*, 17 April 2017.

Louise Hickman, 'The Nature of Self and the Contemplation of Nature: Ecotheology and the History of the Soul', in Michael Fuller (ed.), 'The Concept of the Soul: Scientific and Religious Perspectives' (Cambridge Scholars Publishing, Cambridge, 2014).

Peter Singer, 'Factory Farming: A Moral Issue', *Minnesota Daily*, 22 March 2006.

Yuval Noah Harari, 'Industrial Farming is One of the Worst Crimes in History', *Guardian*, 25 September 2015.

Jonathan Safran Foer, *Eating Animals* (Penguin, London, 2011).

Daniel Imhoff (ed.), *The CAFO Reader: The Tragedy of Industrial Animal Factories* (Watershed Media, New York, 2010).

Jeff Tietz, 'Boss Hog: The Rapid Rise of Industrial Swine', published first in *Rolling Stone*, December, 2006 and reprinted in Daniel Imhoff (ed.), *The CAFO Reader* (Watershed Media, New York, 2010).

Fiona Harvey, Andrew Wasley, Madlen Davies, David Child, 'Rise of the mega farms: how the US model of intensive farming is invading the world', *Guardian*, 18 July 2017.

Lori Marino, Christina M. Colvin, 'Thinking Pigs: A Comparative Review of Cognition, Emotion, and Personality in Sus domesticus', *International Journal of Comparative Psychology*, 28 (23859), 2015.

Lori Marino, 'Thinking Chickens: A Review of Cognition, Emotion and Behavior in the domestic chicken', *Animal Cognition*, 20 (2), 2017.

Áine Caine, 'The meat industry is hiding a dark secret, as workers at "America's worst job" wade through seas of blood, guts, and grease', *Business Insider*, 4 January 2018.

Michael Grabell, 'Exploitation and Abuse at the Chicken Plant: Case Farms built its business by recruiting immigrant workers from Guatemala, who endure conditions few Americans would put up with', *New Yorker*, 1 May 2017.

Lisa Kemmerer, 'Broilers', *Satya* (June/July), 2006.

Upton Sinclair, *The Jungle* (Penguin Books, London, 2002).

Dorothee Brantz, 'Recollecting the Slaughterhouse', Cabinet 4, 2001.

William Cronon, *Nature's Metropolis: Chicago and the Great West* (Norton, New York, 1992).

Christopher Otter, 'Civilizing Slaughter: The Development of the British Public Abattoir, 1850–1910', Brepols Online.

*Le Sang des Bêt*es, a film written and directed by Georges Franju, 1949.

Patrick Modiano, *The Search Warr*ant (Harvill, London, 2002).

Georges Bataille wrote about slaughterhouses in the surrealist magazine *Documents*, the fifteen issues of which he edited from 1929–1930.

Robinson Jeffers, *The Double Axe* (Random House, 1948).

James Karman, *Robinson Jeffers: Poet and Prophet* (Stanford University Press, Stanford, 2015).

Amy J. Fitzgerald, 'A Social History of the Slaughterhouse: From Inception to Contemporary Implications', *Research in Human Ecology*, 17 (1), 2010.

Barbara Pym, *Jane and Prudence* (Virago Modern Classics, London, 2007).

Carol Adams, *The Sexual Politics of Meat* (Bloomsbury Revelations, London, 2015).

Elena Ferrante, *Those Who Leave and Those Who Stay*, Book 3 of the Neapolitan Novels (Europa Editions, New York, 2014).

Arnon Grunberg, *Slaughterhouse*, *Granta* 142, Animalia, February 2018.

Louise Gray, *The Ethical Carnivore* (Bloomsbury Natural History, London, 2017).

Eileen Myles, 'That Rat's Death', from *I Must Be Living Twice* (Tusker Rock Editions, London, 2016).

H. G. Wells, *Anne Veronica* (Penguin Classics, 2005).

Bertha Brupbacher-Bircher, *Health-Giving Dishes*, foreword by Florence Petty, Hon. M.C.A. ('The Pudding Lady') (Edward Arnold, London, 1934).

Adolf Just, *The Jungborn Dietary: A New Vegetarian Cookery Book*, 1905 (mentioned in *The History Cook*, Polly Russell, *F. T.* Magazine, February 2018).

Arnold Arluke and Boria Sax, 'Understanding Nazi Animal Protection and the Holocaust', *Anthrozoös*, 5 (1), 1992.

Dorothée Brantz, 'Stunning bodies: Animal slaughter, Judaism, and the meaning of humanity', *Central European History*, 35 (2), 2002.

Richard Moore-Colyer, Philip Cornford, 'A "Secret Society"? The Internal and External Relations of the Kinship in Husbandry, 1941–52', Cambridge University Press online, 2004 DOI: https://doi.org/10.1017/S09567933030011

Richard Moore-Colyer, 'Rolf Gardiner, English Patriot and the Council for the Church and Countryside', *Agricultural History Review*, 49 (2), 2001.

Val Plumwood, *Feminism and the Mastery of Nature* (Routledge, London and New York, 1993).

Val Plumwood, 'Being Prey', *Terra Nova* 1.3, 1996.

Val Plumwood, 'Integrating Ethical Frameworks for Animals, Humans, and Nature: A Critical Feminist Eco-Socialist Analysis', *Ethics and the Environment* 5 (2), Autumn, 2000.

Val Plumwood, *Environmental Culture: The Ecological Crisis of Reason* (Routledge, London and New York, 2002).

Mary Douglas, *Purity and Danger: An analysis of concepts of pollution and taboo* (Routledge, New York, 2002).

Tracy Warkentin, 'Must Every Animal Studies Scholar Be a Vegan?', The Hypatia Symposium, Hypatia, 27.3, 2012.

Andrew Chignell, Terence Cuneo & Matthew C. Halteman (eds.), *Philosophy Comes to Dinner: Arguments about the Ethics of Eating* (Routledge, New York, 2017).

'Halal or Zabihah?', Islamic Services of America, isahalal.com

'What is Zabihah?', islamawareness.net

Richard Schwartz, 'Judaism and Animals: Vegetarianism', chai-online.org

Gershom Scholem, *Major Trends in Jewish Mysticism* (Schocken Books, New York, 1995).

Herbert Weiner, *9½ Mystics: The Kabbala Today* (Collier Books, New York, 1991).

Perle Epstein, *Kabbalah: The Way of the Jewish Mystic* (Shambhala, Boston and London, 1988).

Perle Besserman, *Kabbalah and Jewish Mysticism* (Shambhala, Boston and London 1997).

5. RIGHTS

The archives of the Wise Club, also known as the Aberdeen Philosophical Society, are in the University of Aberdeen's Special Collections.

Paul B. Wood, *The Aberdeen Enlightenment: The Arts Curriculum in the Eighteenth Century* (Aberdeen University Press, Aberdeen, 1993).

William Paton, *Man and Mouse, Animals in Medical Research* (Oxford University Press, Oxford, 1993).

Miriam Rothschild, *Animals and Man* (Clarendon Press, Oxford, 1986).

David Hume, 'Of the Reason of Animals', from *A Treatise of Human Nature*, III (xvi) (Penguin Classics, London, 1985).

Nathaniel Wolloch, 'The Statues of Animals in Scottish Enlightenment Philosophy', *Journal of Scottish Philosophy*, 4, 2006.

Alejandra Mancilla, 'Nonhuman Animals in Adam Smith's Moral Theory', *Between the Species*, IX, August 2009.

James Burnett Monboddo, *Orang Utans and the Origins of Human Nature* (Continuum Publishing Group, London, 1990).

Comte de Buffon, *Natural History*, Vol. 1–10, Kindle Edition, Amazon Digital Services, 2015, trans. William Smellie, who comments thus on beavers: 'In this society, however numerous, an universal peace is maintained. Their union is cemented by common labours; and it is perpetuated by mutual conveniency, and the abundance of provisions which they amass and consume together. A simple taste, moderate appetites, and an aversion to blood and carnage, render them destitute of the idea of rapine and war.' Smellie is buried in Greyfriars Kirkyard in Edinburgh.

Adam Smith, *Wealth of Nations* (Penguin Classics, London, 1982 & 1999).

Aaron Garrett, 'Francis Hutcheson and the Origin of Animal Rights', *Journal of the History of Philosophy*, John Hopkins University Press, 45 (2), April 2007.

Eberhard Frey, 'The Earliest Medical Texts', Clio Medica, 1985.

Andrew H. Gordon, Calvin W. Schwabe, *The Quick and the Dead: Biomedical Theory in Ancient Egypt (Egyptological Memoirs)*, (Brill, Leiden, Boston, 2004).

Joost J. van Middendorp, Gonzalo M. Sanchez et al., 'The Edwin Smith Papyrus: a clinical reappraisal of the oldest known document on spinal injuries', *European Spine Journal*, 19 (11), November 2010.

Sanib Kumar Ghosh, 'Human cadaveric dissection: a historical account from ancient Greece to the modern era', *Anatomy and Cell Biology*, 48 (3), September 2015.

H. Von Staden, 'The discovery of the body: human dissection and its cultural context in ancient Greece', *Yale Journal of Biological Medicine*, 65 (3), May–Jun 1992.

Noel Si-Yang Bay, Boon-Huat Bay, 'Greek Anatomist Herophilus: the father of anatomy', *Anatomy and Cell Biology*, 43 (4), December 2010.

Rhoda Wynn, 'Saints and Sinners, Women and the Practice of Medicine Throughout the Ages', JAMA, 283 (5), 2000.

Kate Campbell Hurd-Mead, *A History of Women in Medicine: From the Earliest Time to the Beginning of the Nineteenth Century* (Haddam Press, Conn., 1938); onlinebooks.library. upenn.edu

Celsus, *De Medicina*, first published by Nicolaus Laurentii, Alamanus, in Florence in 1478.

At least some of Aelius Galen's work in anatomy is recorded *in Usefulness of the Parts of the Body* written between 165 and 175 AD.

Alexandra Mavrodi, George Paraskevas, 'Mondino de Luzzi: a luminous figure in the darkness of the Middle Ages', *Croatian Journal of Medicine*, 55 (1), 2014.

It was in a letter to fellow Dane, theologian and physician Thomas Bartholin that Nicolas Steno wrote in 1661 to suggest that he found vivisection constituted 'torture'.

W. D. Hall, M.D., 'Stephen Hales: Theologian, Botanist, Physiologist, Discoverer of Hemodynamics', Profiles in Cardiology, Clinical Cardiology 10, 1987.

Pope's words are quoted in *The Works of Alexander Pope*, J. Wharton (ed.), 1797.

The macabre incident of Harvey's post-mortem on his father is referred to in Thomas Wright's book *Circulation: William Harvey's Revolutionary Idea* (Chatto and Windus, London, 2012).

Aubrey's account of his discussion with Harvey: 'he bid me goe to the fountain head and read Aristotle, Cicero and Avicenna, and did call the neotiques shitt breeches', appears in John Aubrey, *Brief Lives* (Vintage Classic, London, 2016).

Christopher Wren gave an account of the experiment in an undated letter to William Petty, probably written during the later 1650s.

Robert Hooke's disquiet was expressed in a letter to Robert Boyle written on 10 November 1664.

Steven Nadler, 'Why Spinoza Was Excommunicated', *Humanities*, 34 (5), September/October, 2013.

In a letter to the *Morning Chronicle* in 1825, Jeremy Bentham wrote: 'I have never seen, nor ever can see, any objection to the putting of dogs and other inferior creatures to pain, in the way of medical experiment, when that experiment has a determinate object, beneficial to mankind, accompanied with a fair prospect of the accomplishment of it.'

Nathaniel Wolloch, 'Rousseau and the Love of Animals', *Philosophy and Literature*, Johns Hopkins University Press, 32 (2), October 2008.

Animals – Voltaire, *The Philosophical Dictionary*, selected and translated by H. I. Woolf, with addenda by J. Perry, 2001 (Knopf, New York, 1924).

Richard Martin lived for much of his life at Ballynahinch Castle, now a hotel. Described in their publicity as 'international father of animal rights, celebrity duellist and rolling financial disaster', his life is celebrated by a one man show named 'Humanity Dick, a Tale of Beasts and Bullets'.

Charles Bell wrote to his brother about his feelings on 1 July 1822, 'Letters of Sir Charles Bell: selected from his correspondence with his brother George Joseph Bell' by the Royal College of Physicians of London, n.80046799.

Darwin and Vivisection, the Darwin Correspondence Project, Cambridge University, Darwin@lib.cam.ac.ukdarwin@lib.cam.ac.uk

Maureen O'Connor, 'Frances Power Cobbe and the Patriarchs', in James H. Murphy (ed.), *Evangelicals and Catholics in Nineteenth Century Ireland* (Four Courts Press, Dublin, 2005).

Moshe Feinsod, Rambam Maimonides, 'Moritz Schiff (1823–1896): A Physiologist in Exile', *Journal of Medicine* 2 (4), October 2011.

Myron Schulz, 'Rudolf Virchow', *Emerging Infectious Diseases*, 14 (9), September 2008.

Anita Guerrini, 'The Rhetorics of Animal Rights', published in *Applied Ethics in Animal Research, Philosophy, Regulation, and Laboratory Applications*, eds. John P. Gluck et al. (Purdue University Press, Indiana, 2002).

Henry Salt, *Animals' Rights: Considered in Relation to Social Progress* (Centaur Press, New York, Society for Animal Rights, Clarks Summit, PA. 1980).

John Oswald's pamphlet was published in 1791 as: 'The Cry of Nature; Or, An appeal to mercy and to Justice on Behalf of the Persecuted Animals by John Oswald, Member of the Club des Jacobines', with an epigraph from Juvenal's *Satires*, XV. ver. 131:

> Mollissima corda
> Humano generi dare se natura satetur
> quae lacrymas dedit: hæe nostri pars optima sensus.
>
> *By giving tears*
> *to the human race Nature revealed she was giving us also*
> *tender hearts; compassion is the finest part of our make-up*

Rachel Rogers, 'The Society of the Friends of the Rights of Man, 1792–94: British and Irish Radical Conjunctions in Republican Paris', La Révolution française [Online], 11 2016: http://journals.openedition.org/lrf/1629 ; DOI: 10.4000/lrf.1629

Martin E. Seligman, G. Beagley, 'Learned helplessness in the rat', *Journal of Comparative and Physiological Psychology*, 88 (2), 1975.

Rami Yankelevitch-Yahav et al., 'The Forced Swim Test as a Model of Depressive-like Behaviour', *Journal of Visualized Experiments* (97), 2015.

The Animal as Sentient Being: The Implications of EU Directive 86/609, ALN Animal Research, 2010.

Brandon Klein, 'The Unappreciated Brilliance of Rats', *Conservation Magazine*, University of Washington, 12 October 2016.

Marc Bekoff, 'Drowning Rats and Human Depression: Positive Psychology for Whom?', *Psychology Today*, January 2012.

Nuno Henrique Franco, 'Animal Experiments in Biomedical Research: A Historical Perspective', *Animals* (Basel), 3 (1), March 2013.

Alison Abbott, 'Neuroscience: The rat pack', *Nature*, 465, 2010.

Akane Nagano, Kenjiro Aoyama, 'Tool-use by rats (*Rattus norvegicus*): tool-choice based on tool features', *Animal Cognition*, 20, September 2016.

D. Panoz Brown et al., 'Rats Remember Items in Context Using Episodic Memory', *Current Biology*, 26 (20), 2016.

Harry F. Harlow, 'The Nature of Love', address of the president at the sixty-sixth Annual Convention of the American Psychological Association, Washington, DC, 31 August 1958. First published in *American Psychologist*, 13, 1958.

Loren Eiseley wrote about the selection of the dog to be used for vivisection in his book *All the Strange Hours: The Excavation of a Life* (Charles Scribner's Sons, New York, 1975).

Richard D. Ryder describes his writing of the 1970 leaflet on speciesism in 'Speciesism Again: The Original Leaflet', published in *Critical Society*, Issue 2, Spring 2010.

Lucius Caviola, Jim A. C. Everett, Nadira S. Faber, 'The Moral Standing of Animals: Towards a Psychology of Speciesism', *Journal of Personality and Social Psychology*, 116 (6), June 2019.

National Centre for Replacement, Refinement and Reduction of Animals in Research, www.nc3rs. org,uk/the-3rs

Kate Connolly, 'VW condemned for testing diesel fumes on humans and monkeys', *Guardian*, 29 January 2018.

Tom Regan, 'The Case for Animal Rights', in Peter Singer, *In Defence of Animals* (Blackwell, New York, 1985).

M. J. Prescott, 'Ethics of Primate Use', Advances in Science and Research, 5, 12 November 2017, pp.11–22.

'The Cambridge Declaration on Consciousness' written by Philip Low, edited by Jaak Panksepp, Diana Reiss, David Edelman, Bruno Van Swinderen, Philip Low and Christof Koch, was presented on 7 July

2012 at the Francis Crick Memorial Conference on Consciousness in Human and non-Human animals at Churchill College Cambridge, and signed by the conference participants in the presence of Stephen Hawking. Among its conclusions, it states: 'Consequently, the weight of evidence indicates that humans are not unique in possessing the neurological substrates that generate consciousness.'

David Ferrier – pioneering neurophysiologist whose research illuminated the role of the cerebral cortex in motor function and epilepsy. Born in Aberdeen, Scotland, 1843, he died in London, England on 28 March 1928, aged eighty-five. Rebecca Akkermans, *Lancet Neurology* (15), June 2016.

J. M. S. Pearce, 'Sir David Ferrier MD, FRS', *Journal of Neurology, Neurosurgery and Psychiatry*, 74 (7), June 2003.

Meigs L. Smimova L. et al., 'Animal testing and its alternatives – the most important omics is economics', Altex, 35, 2018.

Matta Busby, 'Barbaric tests on monkeys lead to calls for closure of German lab', *Guardian*, 15 October 2019.

The Jane Goodall Institute, October 2019; the janegoodallintstitue.com

Luechtefeld T., Marsh, D. et al., 'Machine Learning of Toxicological Big Data Enables Read-Across Structure Activity Relationships (RASAR) Outperforming Animal Test Reproducibility', *Toxicological Sciences*, 165 (1), 1 September 2018.

Kathleen Baird Murray, 'Testing times for animal welfare', *Financial Times*, 29 September 2016.

Kirk R. Wilhelmus, 'The Draize Eye Test', Survey of Opthalmology, 45 (6) May–June 2001.

Shripriya Singh, Vinay K. Khanna, Aditya B. Pant, 'Development of In Vitro Toxicology: A Historic Story', Chapter One, *In Vitro Toxicology*, Alok Dhawan & Seok Kwon (eds.) (Academic Press, Cambridge, Mass., 2018).

6. THE MUSEUM

Alison K. Brown, 'The Kelvingrove "New Century" Project: Changing Approaches to Displaying World Cultures in Glasgow', *Journal of Museum Ethnography*, No.18, May 2006.

Henry Farmer Papers, Glasgow University Library Special Collections Department.

Paul J. Crutzen, 'The Anthropocene', in E. Ehlers, T. Krafft (eds.), *Earth System Science in the Anthropocene* (Springer, Berlin, Heidelberg, 2006).

Patricia A. Ganea et al., 'Do cavies talk? The effects of anthropomorphic picture books on children's knowledge about animals', *Frontiers in Psychology*, 5 (23), 2014.

Marla V. Anderson, Antonia Henderson, 'The Looking Glass: New Perspectives on Children's Literature', lib.latrobe.edu.au

Anderson, M. V. and Henderson, A. J. Z., 'Pernicious Portrayals: The Impact of Children's Attachment to Animals of Fiction on Animals of Fact', *Society and Animals* 13 (4), December 2005.

Selma G. Lanes, 'Male Chauvinist Rabbits', *New York Times*, 30 June 1974.

Kadri Tuur, Morten Tonnessen (eds.), *The Semiotics of Animal Representations* (Nature, Culture and Literature) (Rodopi, Amsterdam, 2014).

George Boas, *The Happy Beast in French Thought of the Seventeenth Century* (Octagon Books, New York, 1933).

Bruta animalia ratione uti – Gryllus, Plutarch, penelope.uchicago.edu, as published in Vol. XII of the Loeb Classical Library edition, 1957.

Yannick Joye, Andreas de Block, '"Nature and I Are Two": A Critical Examination of the Biophilia Hypothesis', *Environmental Values*, 20 (2), May 2011.

Ryan Gunderson, 'The First-generation Frankfurt School on the Animal Question: Foundations for a Normative Sociological Animal Studies', *Sociological Perspectives*, 57 (4), April 2014.

Theodor Adorno composed a song of welcome to Max Horkheimer – 'Rüsselmammuts Heimkehr'.

Herbert Marcuse's essay 'Ecology and Revolution' was read at a symposium held in Paris in 1972. Extracts were later published in *Liberation* No.16, September 1972.

Erich Fromm, 'Creators and Destroyers', *Saturday Review*, New York, 1964.

Fromm's Credo, fromm-online.org

Theodor Adorno, *Minima Moralia: Reflections on a Damaged Life* (Verso Books, New York, 2005).

Max Horkheimer wrote to the Californian congressman Ned R. Healy in 1945 to ask him to help pass a bill to outlaw the use of dogs in vivisection.

'The Skyscraper' was one of the short reflections which appeared in Max Horkheimer's *Dawn and Decline, Notes 1926–1931 and 1950–1969* (Seabury Press, New York, 1978).

J. D. Salinger, *The Catcher in the Rye* (Penguin, London, 1994).

Alexandra Alvis, 'Elizabeth Gould: An Accomplished Woman', Biodiversity Heritage Library, originally posted on the Smithsonian library blog, 29 March 2018.

Frances Stonor Saunders, 'How to Be a Stuffed Animal', *Lapham's Quarterly*, Animals, Vol. VI (2), Spring 2013.

Olga Nieuwenhuys, 'Can the Teddy Bear Speak', *Childhood*, 18 (4), 18 November 2011.

Memorandum submitted by Glasgow City Council on repatriation requests: https://publications.parliament.uk/pa/cm199900/cmselect/cmcumeds/371/0051808.htm

The Brussels Universal Exhibition in 1958 was designed to celebrate and justify colonialism and featured Congolese artists displaying and selling their work. Shortly after the exhibition closed, the Congolese National Movement was established by Patrice Lumumba, heralding the end of Belgian colonial rule.

An exhibition in Yvoir in Belgium in 2002 featured an artificial 'pygmy village' where Cameroonians who were brought to Belgium for the occasion performed songs and dances. It used the slogan 'Discover Black Africa and the pygmies' in its publicity. In June 2005, Augsburg Zoo in Germany held a similar event.

Caroline Goldstein, 'How Many Animals Have Died for Damien Hirst's Art to Live? We Counted. Nearly one million, by our conservative estimate', *Artnet News*, 13 April 2017.

The David Winton Bell Gallery at Brown University in Providence, Rhode Island, staged an exhibition in 2016–2017 – 'Dead Animals, or the curious occurrence of taxidermy in contemporary art'.

Roger Lovegrove, *Silent Fields: The Long Decline of a Nation's Wildlife* (Oxford University Press, Oxford, 2008).

Philip Hoare, 'Take down your taxidermy – it's a disgrace', *Guardian*, 6 August 2015.

B'revach ben olam v'za'azuah: al ha model b'tarbut haisraelit, 'In the Space Between World and Toy: The Model in Israeli Culture', Tami Berger (Resling, Tel Aviv, 2008).

'Harvey Nichols ran a tongue in cheek campaign to celebrate the store's rare jewellery collection. The ad features a Northern Bald Ibis, which is considered critically endangered, wearing an equally rare tanzanite necklace', *Campaign Magazine*, 9 May 2017.

The Northern Bald Ibis advert was devised by Henry Westcott and Eduardo Balestra of Adam & Eve DDB.

Audrey Niffenegger, 'A Few of My Favourite Things', *New Statesman*, 27 July–9 August 2018.

Lidija M. McKnight, Stephanie D. Atherton-Woolham, Judith E. Adams, 'Imaging of Ancient Egyptian Animal Mummies', *Radio Graphics*, 35 (7), November 2015.

Richard Sutcliffe, Martin Rutherford, Jeanne Robinson, 'Sir Roger the Elephant' in *The Afterlives of Animals: A Museum Menagerie*, Samuel J. M. M. Alberti (ed.) (University of Virginia Press, Charlottesville, 2013).

7. TRADITION

'What is Tradition?', Nelson H. H. Graburn, *Museum Anthropology*, 24 (2/3), 2001.

'Saving the whale is no longer just a slogan', a letter from Chris Butler-Stroud, Chief Executive, Whale and Dolphin Conservation, to the *Financial Times*, 22 December 2018.

'The History of Whaling and the International Whaling Commission', wwf.panda.org

'A Brief History of whales and commercial whaling', greenpeace.org.uk

'Start of Whaling, National Museum of Australia', www.nma.gov.au/ defining-moments

Adam Wernick, 'How pop music helped save the whales', Studio 360, 22 December 2014.

Jo Roman et al., 'Whales as marine ecosystems engineers', *Frontiers in Ecology and the Environment*, 12 (7), 3 July 2014.

Justin McCurry, 'Japan resumes commercial whaling for first time in 30 years', *Guardian*, 1 July 2019.

The North Atlantic Marine Mammal Commission: nammco.no – Hunting Manuals: Use and Maintenance of Weapons and Gears in Whaling. There are manuals on pilot whaling, small whale hunting and the hunting of baleen whales.

Chris Burgess, '"Killing the Practice of Whale Hunting is the same as Killing the Japanese People": Identity, National Pride, and Nationalism in Japan's Resistance to International Pressure to Curb Whaling', *Asia-Pacific Journal*, 14 (8/2), 15 April 2016.

Luke Davis, 'On the Supposed Importance of Cultural Traditions for Whaling Practice', *Oxford Ethics*, 15 September 2014.

Ralph Chami et al., 'Nature's Solution to Climate Change: A strategy to protect whales can limit greenhouse gases and global warming', IMF Finance and Development 56 (4), December 2019.

'Norwegians Hate Whale Meat. So Why Does the Country Insist on Whaling?', Brian Palmer, HuffPost, 9 October 2014, updated 6 December 2017.

Pal Weihe, Høgni Debes Joensen, 'Toxic Menu; Database with levels of toxic substances in whale meat', www.toxic-menu.org

'Dietary recommendations regarding pilot whale meat and blubber in the Faroe Islands', *International Journal of Circumpolar Health*, 71 (10), 10 July 2012.

Benedict Esmond Singleton, Russell Fielding, 'Inclusive Hunting: examining Faroese whaling using the theory of socio-cultural viability', *Maritime Studies*, 16 (6) December 2017.

J. P. Desforges et al., 'Predicting global killer whale population collapse from PCB pollution', *Science*, 361 (6409), 28 September 2018.

Sarah Zielinski, 'Whales are full of toxic chemicals', *Science News*, 19 January 2016.

Jurriaan M. de Vos et al., 'Estimating the normal background rate of species extinction', *Conservation Biology*, 29 (2), 26 August 2014.

Leonardo da Vinci – ucmp.berkeley.edu.history.evolution

'Commission refers Finland to Court over spring hunting of wild birds', European Commission, 8 March 2018 (MEMO/18/1444).

Meilan Solly, 'Ortolans, Songbirds Enjoyed as French Delicacy, Are Being Eaten Into Extinction', Smithsonian.com, 23 May 2019.

F. Jiguet et al., 'Migration Strategy of the Ortolan Bunting: Final Report of the Scientific Committee', Office national de la chasse et de la faune sauvage, Paris, 2016.

From: Anthony Bourdain, *Medium Raw: A Bloody Valentine to the World of Food and the People Who Cook* (Bloomsbury, London, 2010): 'I bring my molars down and through the bird's rib cage with a wet crunch and am rewarded with a scalding rush of burning fat and guts down my throat. Rarely have pain and delight combined so well.'

Teresa Lappe-Osthege, 'Endangered British birds to be hunted under new permit – here's how that could fuel an illegal pan-European trade', *The Conversation*, 25 January 2019.

Benjamin Barca et al., 'Environmentalism in the crosshairs: Perspectives on migratory bird hunting and poaching conflicts in Italy', *Global Ecology and Conservation*, 6 April 2016.

Stuart Butchart, 'Eight million birds killed illegally at 20 Mediterranean locations each year', BirdLife, 4 March 2016.

'An Open Hunting Season on the Red-listed Turtle Dove', BirdLife Malta, 17 May 2019.

BirdLife International and the Committee Against Bird Slaughter (CABS) are valuable sources of information on all aspects of bird hunting and trapping.

Brian Chang et al., 'How seabirds plunge-dive without injuries', PNAS 113 (43), 25 October 2016.

Wildlife and Countryside Act, 1981, legislation.gov.uk

John Beatty, *Sula: The Seabird-Hunters of Lewis* (Michael Joseph, London, 1992).

Donald Murray, *The Guga Hunters* (Birlinn, Edinburgh, 2015).

'Turning to the issue of animal welfare. It is the Scottish Government's understanding that most of the gugas will be killed by a single blow to the head. Where a second blow is required, it is very likely that the first will have rendered the bird unconscious. In our view therefore the method used to kill the gugas does not involve unnecessary suffering. Given the above and that the guga hunt is carried out in accordance with a licence issued by the Scottish Government we are confident that the guga hunt is compatible with the requirements of section 19 of the Animal Health and Welfare (Scotland) Act 2006, which allows the killing of an animal in an appropriate and humane manner . . .' Jonathan Young, Policy Officer, Wildlife Management Team, The Scottish Government, September 2010.

From Robert Macfarlane's *The Old Ways* (Penguin, London, 2013): 'I thought of how, once the guga-hunting party had departed from Sula Sgeir each year, the amputated wings of the dead gannets – 4,000 wings from 2,000 birds – were left lying on the summit, so that when the next big autumn storm came and the next big wind blew from the south or the west, thousands of these severed wings would lift from the surfaces of the island, such that it seemed, when seen from the sea, that the rock itself were trying to lift off in flight – an entire island rising into the air, like Swift's Laputa.'

The Custom of Kapparot in the Jewish Tradition, The Jewish Vegetarian Society, www.jvs.org.uk

Gerardo Ceballos, Paul R. Ehrlich et al., 'Accelerated modern human-induced species losses: Entering the sixth mass extinction', *Science Advances*, 1(5), 19 June 2015.

Jonathan Watts, 'Red list research finds 26,000 global species under extinction threat', *Guardian*, 5 July 2018.

Damian Carrington, 'Humans just 0.01% of all life but have destroyed 83% of wild mammals', *Guardian*, 21 May 2018.

Yinon M. Bar-On, Rob Phillips, and Ron Milo, 'The biomass distribution on Earth', PNAS, 115 (25), 19 June 2018.

Matt Davis, Søren Faurby, Jens-Christian Svenning, 'Mammal diversity will take millions of years to recover from the current biodiversity crisis', PNAS, 115 (44), 30 October 2018.

Ralph C. Croizier, 'Traditional Medicine in Communist China: Science, Communism and Cultural Nationalism', *The China Quarterly*, No.23, July–September 1965.

Jessica Murray, 'World's donkeys being "decimated" by demand for Chinese medicine', *Guardian*, 21 November 2019.

Susan A. Mainka and Judy A. Mills, 'Wildlife and Traditional Chinese Medicine: Supply and Demand for Wildlife Species', *Journal of Zoo and Wildlife Medicine*, 26 (2), 1995.

Judith Shapiro, *Mao's War Against Nature: Politics and the Environment in Revolutionary China* (Cambridge University Press, Cambridge, 2011).

J. Still, 'Use of animal products in traditional Chinese medicine: Environmental impact and health hazards', *Complementary Therapies in Medicine*, 11 (2), July 2003.

Miriam Gross, 'Between Party, People, and Profession: The Many Faces of the "Doctor" during the Cultural Revolution', *Medical History*, 62 (3), July 2018.

Henry Mance, 'The Great Emptying', *Financial Times* Magazine, 23/24 November 2019.

8. THE HUNT

'Those who hunt for fun and games are like "madmen shooting flaming arrows of death"', Shabbetai Elhanan ben Elisha del Vecchio, born 1707.

The Jewish view of hunting is expressed in Rabbi Yechezkel Landau's *Responsa Noda be Yehuda II Yoreh Deah 10.*

'Hunting', *Jewish Encylopedia*, Vol. VI (Funk and Wagnalls, New York, 1925).

Annie Dillard, *The Writing Life* (HarperPerennial, New York, 1989).

Nadine Weidman, 'Popularizing the Ancestry of Man: Robert Ardrey and the Killer Instinct', *Isis*, 102 (2), June 2011.

'Caton-Thompson, Women's Networks, and Racial Politics in Great Zimbabwe', Lady Science 2016; Ladyscience.com

Ute Deichmann, *Biologists Under Hitler* (Harvard University Press, Cambridge, Mass., London, 1996).

Jan Dizard, 'Hunting for a Sustainable Relationship With Nature', Center for Humans and Nature, 3 March 2014.

Jan Dizard, *Mortal Stakes: Hunters and Hunting in Contemporary America* (University of Massachusetts Press, Amherst, 2003).

Anne Fadiman, *Ex Libris: Confessions of a Common Reader* (Penguin, London, 1999).

Marti Kheel, *Nature Ethics: An Ecofeminist Perspective* (Rowman & Littlefield, Maryland, 2007).

Marti Kheel, 'License to Kill: An Ecofeminist Critique of Hunters' Discourse', in Carol J. Adams and Josephine Donovan (eds.), *Animals and Women: Feminist Theoretical Explorations* (Duke University Press, Durham, NC, 1995), pp.85–125.

Brian Luke, 'Violent love: Hunting, heterosexuality, and the erotics of men's predation', *Feminist Studies*, 24 (3), 1998, pp.627–55.

Brian Luke, 'A Critical Analysis of Hunters' Ethics', *Environmental Ethics*, 19 (1), 1997.

Ted Kerasote, *Bloodties: Nature, Culture, and the Hunt* (Kodansha, San Francisco, 1993).

Giles Catchpole, *Birds, Boots and Barrels: Game Shooting in the 21st Century* (Swanhill Press, Shrewsbury, 2002).

'A Precision-Guided Firearm (PGF) is a comprehensive, purpose-built weapon system that leverages the same tracking and fire-control technology found in advanced fighter jets. The TrackingPoint PGF system is the first and only rifle optics system to offer the advanced technology that guides the release of ordnance. Known as TriggerLink, this fire control system virtually eliminates human error caused by misaiming, mistiming, and central nervous system jitter. With a TrackingPoint system, hunters unable to dedicate the time and resources required to achieve such an elite level of skill can now make dreams of a trophy kill a reality.' Tracking-point.com

Gillian Tett, '"Smart guns" and the dangers of trigger-happy technology: A new generation of high-tech rifles brings home the terrifying pace of digital disruption', *Financial Times* Magazine, 6/7 April 2019.

Joy Williams, 'The Killing Game: Why the American hunter is bloodthirsty, piggish, and grossly incompetent', *Esquire* Magazine, October 1990.

Susan Griffin, *Women and Nature: The Roaring Inside Her* (Counterpoint, Berkeley, 2016).

Pam Houston (ed.), *Women on Hunting* (Ecco Press, New Jersey, 1995).

Adventuress Magazine, tailored for the Outdoor Women: betheadventuress.com

Wildlife Enthusiast Magazine, 'For Women Who Love Hunting, Fishing and the Great Outdoors'; wildlifeenthusiast.com

Henry Salt, 'The Blooding of Children', from *Killing for Sport* (George Bell and Son, London, 1914).

Patrick Durkin, 'Post-Kill Rituals: Matters of the Heart', NRA American Hunter, 14 August 2016.

Boone and Crockett Club: www.boone-crockett.org

Crossbow Revolution Magazine: published by Game and Fish.

Chelsea Batavia, Michel Paul Nelson et al., 'The Elephant (Head) in the Room: A critical look at trophy hunting', Conservation Letters, 12 (1), January/February 2019.

Andrew Loveridge, *Lion Hearted: The Life and Death of Cecil and the Future of Africa's Iconic Cats* (Regan Arts, 2018).

Chris F. Darimont et al., 'Why Men Trophy Hunt', *Biology Letters*, 13 (3), 19 March 2017.

10 Facts About Trophy Hunting: www.bornfree.org/articles/trophy-hunting-facts.

10 Facts About Canned Hunting: www.bornfree.org/articles/canned-hunting-facts.

Jonathan Leake, 'Britons flock to package-trip trophy hunts', *Sunday Times*, 29 September 2019.

John R. Allen, 'Bloodsports on Deeside', from *Grampian Hairst, An Anthology of Northeast Prose*, William Donaldson, Douglas Young (eds.) (Aberdeen University Press, 1981).

Adam Watson, Jeremy D. Wilson, 'Seven decades of mountain hare counts show severe declines where high-yield recreational game bird hunting is practised', *Journal of Applied Ecology*, 55, 4 April 2018. (One of the authors of this paper, the distinguished biologist Adam Watson, expert on so much connected with Scotland's natural world, in particular ptarmigan and golden eagles, died aged eighty-eight in January 2019.)

Megan Murgatroyd, Stephen M. Redpath et al., 'Patterns of satellite tagged hen harrier disappearances suggest widespread illegal killing on British grouse moors', *Nature Communications*, 10 (1094), 19 March 2019.

rspb.org.uk 28 August 2018: 'Moorland gamekeeper Timothy Cowin has today pleaded guilty to two charges concerning the intentional killing of two protected short-eared owls on the Whernside Estate in Cumbria, an area managed for driven grouse shooting. He also pleaded guilty to one charge relating to the possession of items capable of being used to commit offences against wild birds.'

Joah R. Madden, Andrew Hall, Mark A. Whiteside, 'Why do many pheasants released in the UK die, and how can we best reduce their natural mortality?', *European Journal of Wildlife Research*, 64 (40), 2018.

Tim M. Blackburn, Kevin J. Gaston, 'Abundance, biomass and energy use of native and alien breeding birds in Britain', *Biological Invasions*, 20, 11 July 2018.

N. J. Aebischer, 'Fifty-year trends in UK hunting bags of birds and mammals, and calibrated estimation of national bag size, using GWCT's National Gamebag Census', *European Journal of Wildlife Research*, 65, 2019.

From James Purdey & Sons: 'A Guide to Gamebirds': 'Every breed of game is different and deserving of our full respect' ... 'The red-legged partridge is native to the Iberian Peninsula, Italy, Corsica and France. Its origins probably explain why it's also called the Frenchman, though its decriers claim it came from its preference for running, unlike the "brave little native" – the grey or English partridge – which more readily faces the Guns. Others maintain it was so christened because its red legs match the crimson trousers worn by French infantry prior to the First World War.' Purdey.com

International Order of St Hubertus: iosh-usa.com

Jeffrey Toobin, 'The Company Scalia Kept', *New Yorker*, 2 March 2016.

9. THE COAT

Julia Emberley, *The Cultural Politics of Fur* (Cornell University Press, New York, 1997).

Schumpeter, 'Adventures in the skin trade', *Economist*, 3 May 2014.

Bridget Foley, 'The Fur Debate', *Women's Wear Daily*, 26 April 2018.

House of Commons Environment, Food and Rural Affairs Committee inquiry on the fur trade in the UK, April 2018. British Fur Trade

Association. Written Evidence to EFRA Select Committee, February 2018: http://data.parliament.uk/writtenevidence/committeeevidence. svc/evidencedocument/environment-food-and-rural-affairs-committee/fur-trade-in-the-uk/written/78933.html

Lyudmila Trut et al., 'Animal evolution during domestication: the domesticated fox as a model', *Bioessays*, 3 (31), March 2009.

Lee Alan Dugatkin, 'The silver fox domestication experiment', *Evolution: Education and Outreach*, 11 (16), 7 December 2018.

Greger Larsen, Dorian Q. Fuller, 'The Evolution of Animal Domestication', *Annual Review of Ecology, Evolution and Systematics*, 45, November 2014.

'Sable's behaviour resembles the of cat's (sic), and although special selective work in order to tame sable hasn't yet been carried out, it is clear, that this is only a matter of time,' International Fur Auction Sojuzpushnina: www.sojuzpushnina.ru/en/c/51/

Tazarve Gharajehdaghipour et al., 'Arctic foxes as ecosystem engineers: increased soil nutrients lead to increased plant productivity on fox dens', *Scientific Reports*, 6 (24020), 5 April 2016.

Claudia Vinke et al., 'To swim or not to swim: An interpretation of farmed mink's motivation for a water bath', *Applied Animal Behaviour Science*, III(1–2), 2008.

Georgia Mason et al., 'Frustrations of fur-farmed mink', *Nature*, 410 (1), 2001.

Rebecca K. Meagher et al., 'Who's afraid of the big bad glove? Testing for fear and its correlates in mink', *Applied Animal Behaviour Science*, 133 (3–4), September 2011.

Rebecca K. Meagher et al., 'Sleeping tight or hiding in fright? The welfare implications of different subtypes of inactivity in mink', *Applied Animal Behaviour Science*, 144 (3–4), March 2013.

Valeria Franchi et al., 'Fur chewing and other abnormal repetitive behaviors in chinchillas (*Chinchilla lanigera*), under commercial fur-farming conditions', *Journal of Veterinary Behaviour*, 11, Jan/Feb 2016.

'Fur Farms Break EU Animal Protection Law': European Interest, 21 November 2019: 'Watching damning undercover footage of appalling conditions on European fur farms, MEPs from Finland,

France, Luxembourg and Poland called fur farming incompatible with EU animal protection law at an event held at the European Parliament by a coalition of animal welfare organisations.'

Kyra Gremmen, 'Safeguarding Animal Welfare in the European Fur Farming Industry', Bachelor Thesis in European Public Administration, University of Twente, 2014.

Lesley A. Peterson, 'Brief Summary of Fur Laws and Fur Production', Animal Legal and Historical Center, 2010.

Fur farming legislation in the EU: 'There is no specific EU legislation providing detailed animal welfare requirements for the keeping of animals for fur production. Fur factory farms are covered by Council Directive 98/58/EC [5], which lays down the general minimum requirements for the protection of all animals kept for farming purposes', Humane Society International, hsi.org.

Welfur Welfare Assessment Protocols for Mink. http://fureurope.eu/wp-content/uploads/2015/10/Mink_ protocol_ nal_web_edition_light.pdf

Welfur Welfare Assessment Protocols for Foxes. http://fureurope.eu/wp-content/uploads/2015/10/WelFur_ fox_protocol_web_edition.pdf

Leo Tolstoy, *Writings on Civil Disobedience and Nonviolence*, 1886.

Luke Andrew, 'Mummified bodies, cannibalism and living in faeces: Shocking footage exposes a French fur farm as animal rights campaigners call for it to be closed down', MailOnline, 3 July 2019.

Ross Ibbotson, 'Rotting foxes and starving dogs are discovered at Polish "fur farm" with bones jutting out of cruelly mistreated animals' corpses', MailOnline, 23 September 2019.

Luke Andrews, 'Horrifying footage shows a cannibalised mink, a decomposing fox and shivering cubs at Finnish fur farms which export millions of pounds worth of pelts to the UK each year', MailOnline, 18 October 2019.

Nada Farhoud, 'Inside horror of fur farms where distressed minks and foxes resort to cannibalism: Anti-fur campaigners have visited 13 farms in Finland that supply the UK market to discover harrowing scenes of animals suffering in poor living conditions', *Mirror*, 16 October 2019.

Heli Nordgren, 'Characterization of a New Epidemic Necrotic

Pyoderma in Fur Animals and its Association with *Arcanobacterium phocae* Infection', Plos 9 (10), 10 October 2017.

Sami A. Jokinen, et al., 'A 1500-year multiproxy record of coastal hypoxia from the northern Baltic Sea indicates unprecedented deoxygenation over the 20th century', *Biogeosciences*, 2018.

Heidrun Fammlet et al., 'The Story of the Eutrophication and Agriculture of the Baltic Sea', May 2018: responseable.eu

Manfred Krautter, Poison in Furs: Report 2011, EcoAid and Four Paws.

'Toxic Fur: A Global Issue', Research in China, ACT Asia, 2018: actasia.org

'Morality is Not a Luxury', Fur Europe Annual Report 2015, p.22.

Mathieu Diribarne, 'Characterization of the gene and mutation responsible for the rex hair trait in the rabbit', Génétique Animale et Biologie Intégrative Unité Mixte de Recherche INRA-AgroParisTech, May 2011.

Sarah Malm, 'Sickening undercover footage shows the horrific cramped conditions rabbits bred for their fur are kept in before being skinned and used for luxury clothing in France', MailOnline, 21 December 2017.

Joseph Goldstein, 'Niche Trade in Lamb Pelts Proves Vital to Ailing Afghan Economy', *New York Times*, 23 April 2015.

'As the first day of the online fur festival – the first day ever dedicated to fur shopping – is approaching in China, Kopenhagen Fur strategically teamed up with Lieer Baby. The Chinese influencer has 2.6 million followers and is the third biggest account on the Chinese streaming site Tao Bao.' Website of Kopenhagen Fur, 8 October 2019.

'Laws on Leg-Hold Animal Traps Around the World', Library of Congress, 2016.

Hannu Korhonen et al., 'A Questionnaire Study on Euthanasia in Farm-Raised Mink', *International Journal for Educational Studies*, 5 (2), 2013.

Republic of Turkey, Ministry of Agriculture and Forestry, Fur Farming: 'Since furs with bullet hole are of poor quality (sic), use of trap or insecticide is recommended for hunting ... Fur farming, meaning to keep animals for their furs, just like keeping animals for their meat, to feed them properly in proper barns in a controlled and limited way in accordance with the demand . . .'

Lucy Bannerman, 'Fur flies as lynx coat attracts wrong look', *The Times*, 12 January 2019: 'The £68,000 lynx fur coat was billed as "the pinnacle of trend-setting bravery".'

10. WHAT IS LOVE?

Francis D. Lazenby, 'Greek and Roman Household Pets', *The Classical Journal*, 44 (4), January 1949.

Eleanor Ainge Roy, 'New Zealand bans swimming with bottlenose dolphins after numbers plunge', *Guardian*, 28 August 2019.

Brett Mills, 'Television wildlife documentaries and animals' right to privacy', *Journal of Media and Cultural Studies*, 24 (2), 26 March 2010.

Penthai Siriwat, Vincent Nijman, 'Illegal pet trade: on social media as an impediment to the conservation of Asian otter species', *Journal of Asia-Pacific Biodiversity*, 11 (4), 1 December 2018.

K. Anne-Isola Nekaris et al., 'Tickled to Death: Analysing Public Perceptions of "Cute" Videos of Threatened Species (Slow Lorises – *Nycticebus spp.*) on Web 2.0 Sites', PLoS One, 8 (7), 24 July 2013.

Theodore Roethke, *The Lizard*, Collected Poems (Faber, London 1966).

'A stroll through the worlds of animals and men: A picture book of invisible worlds', Jakob Von Uexküll, 1934.

Brett Buchanan, *Onto-Ethologies: The Animal Environment of Uexküll, Heidegger, Merleau-Ponty, and Deleuze* (State University of New York Press, Syracuse, 2008).

Jacques Derrida, *The Animal That Therefore I Am*, (ed.) Marie-Louise Mallet, trans. David Wills (Fordham University Press, New York, 2008).

Jacques Derrida, Elisabeth Roudinesco, *For What Tomorrow . . . A Dialogue*, trans. Jeff Fort (Stanford University Press, California, 2004).

Christine Gerhardt, 'Narrating entanglement: Cixous' "Stigmata, or Job the Dog"', Humanities 6 (75), 2017.

Jean Grondin, 'Derrida and the Question of Animal', Cités, 2 (30), 2007.

Zipporah Weisberg, '"The Simple Magic of Life": Phenomenology, Ontology, and Animal Ethics', *Humanimalia*, 7 (1), 2015.

Margaret Renki, 'Praise Song for the Unloved Animals', *New York Times*, 27 May 2019.

Céline Albert et al., 'The Twenty Most Charismatic Species', PLoS ONE, 13 (7), 9 July 2018.

Daniel T. C. Cox, Kevin Gaston, 'The Likeability of Garden Birds: Importance of Species Knowledge and Richness in Connecting People to Nature', PLoS ONE, 10 (11), 11 November 2015.

Gunnthorsdottir, A., 'Physical attractiveness of an animal species as a decision factor for its preservation', *Anthrozoös*, 14 (4), 2001, pp.204–15.

Konrad Lorenz, 'Die angeborenen Formen möglicher Efahrung', *Zeitschrift für Tierpsychologie*, 5, 1943.

Jessika Golle et al., 'Sweet Puppies and Cute Babies: Perceptual Adaptation to Babyfacedness Transfers across Species', PLoS One, March 2013.

Laura R. Botiqué et al., 'Ancient European dog genomes reveal continuity since the Early Neolithic', *Nature Communications*, 8 (16082), 18 July 2017.

Claudio Ottoni et al., 'The palaeogenetics of cat dispersal in the ancient world', *Nature Ecology and Evolution*, 1 (0139), 19 July 2017.

'Genetic Welfare Problems of Companion Animals': Universities Federation for Animal Welfare: An information resource for prospective pet owners: ufaw.org.uk

Niels Pederson et al., 'A genetic assessment of the English bulldog', *Canine Genetics and Epidemiology*, 3 (6), 2016.

D. Gunn-Moore et al., 'Breed-related disorders of cats', *Journal of Small Animal Practice*, 49 (4), April 2008.

Lindsay L. Farrell et al., 'The challenges of pedigree dog health: approaches to combating inherited disease', *Canine Genetic Epidemiology*, 2 (3), 11 February 2015.

Jane Ladlow et al., 'Pedigree Dogs Exposed: Five Common problems in pedigree dogs', RSPCA, Australia.

'Brachycephalic obstructive airway syndrome', *Veterinary Record*, 182 (3), 2018.

Susan Sontag, 'Elias Canetti', The Modern Common Wind, *Granta* 5, 1 March 1982.

Elias Canetti: 'The most beautiful monument to man would be a horse that has thrown its rider.'

Alexandra Horowitz, 'Dogs Are Not Here for Our Convenience', *New York Times*, 3 September 2019.

Lindsay Parenti et al., 'A Revised Taxonomy of Assistance Animals', *Journal of Rehabilitation Research and Development*, 50 (6), 2013.

Leonardo Caffo, 'Animality as a new space paradigma'; comradeanimal. com

Gabriella Airenti, 'The Development of Anthropomorphism in Interaction: Intersubjectivity, Imagination, and Theory of Mind', *Frontiers in Psychology*, 9 (2136), 2018.

Lori Gruen (ed.), *Critical Terms for Animal Studies* (University of Chicago Press, 2018).

Gregory S. Okin, 'Environmental impacts of food consumption by dogs and cats', PLoS, 2 August 2017; https://doi.org/10.1371/journal. pone.0181301

Roxanne D. Hawkins, Joanne M. Williams, and Scottish Society for the Prevention of Cruelty to Animals, 'Childhood Attachment to Pets: Associations between Pet Attachment, Attitudes to Animals, Compassion, and Humane Behaviour', *International Journal of Environmental Research and Public Health*, 14 (5), 6 May 2017.

Michelle Newberry, 'Pets in danger: Exploring the link between domestic violence and animal abuse', *Aggression and Violent Behaviour*, 34, May 2017.

Vivek Upadhya, 'The Abuse of Animals as a Method of Domestic Violence: The Need for Criminalization', *Emory Law Journal*, 17 April 2013.

S. E. Macdonald et al., 'Children's experience of companion animal maltreatment in households characterized by intimate partner violence', *Child Abuse and Neglect*, 50, December 2015.

Alison Hawthorne Deming, *Zoologies: On Animals and the Human Spirit* (Milkweed Editions, Minnesota, 2014).

Charles Siebert, 'What Does a Parrot Know About PTSD?', *New Yorker*, 28 January 2016.

C. S. Lewis, *The Problem of Pain* (Collins, London, 2012).

Barry Yeoman, 'When Animals Grieve', National Wildlife Federation, 30 January 2018.

G. Bearzi et al., 'Whale and dolphin behavioural responses to dead conspecifics', Zoology (Jena), 128, June 2018.

Brigit Katz, 'Unique Brain Circuitry Might Explain Why Parrots Are So Smart', *Smithsonian Magazine*, 8 July 2018.

José Tella, Fernando Hiraldo, 'Illegal and Legal Parrot Trade Shows a Long-Term, Cross-Cultural Preference for the Most Attractive Species Increasing Their Risk of Extinction', PLoS ONE, 9 (9), 16 September 2014.

Tim Knight, Fauna & Flora International, 'Forest fire pushes imperilled parrot closer to the brink', Phys Org, 11 June 2019.

Shannon Sims, 'The Amazon's Blazing Fires Are Squeezing Habitat for Imperiled Birds: The fires are not natural to the ecosystem – and add fuel to mounting threats that birds already face in the region', Audubon.org, 9 September 2019.

Gary L. Francione, Anne E. Charlton, 'The Case Against Pets', aeon.co, 8 September 2016.

Linda Rodriguez McRobbie, 'Should we stop keeping pets? Why more and more ethicists say yes', *Guardian*, 1 August 2017.

Marion Nestle, 'Haggadah: The Passover Meal', foodpolitics.com, 12 March 2012.

Maria Baghramian, 'Ireland in the life of Ludwig Wittgenstein', *Hermathena*, No.144, Summer 1988.

Further Reading

Adams, Carol, *The Sexual Politics of Meat: A Feminist-Vegetarian Critical Theory* (Continuum, New York, 1991).

Adorno, Theodor, Horkheimer, Max, trans. John Cummings, *The Dialectics of Enlightenment* (Verso, London & New York, 1979).

Allen, Barbara, *Pigeon* (Reaktion, London, 2009).

Aloi, Giovanni, *Speculative Taxidermy: Natural History, Animal Surfaces, and Art in the Anthropocene* (Columbia University Press, New York, 2018).

Ames, Eric, *Carl Hagenbeck's Empire of Entertainments* (University of Washington Press, Seattle, 2009).

Avery, Mark, *Inglorious – Conflict in the Uplands* (Bloomsbury Natural History, London, 2015).

Balcombe, Jonathan P., *Pleasurable Kingdom: Animals and the Nature of Feeling Good* (Macmillan, London and New York, 2006).

——, *What a Fish Knows: The Inner Lives of Our Underwater Cousins* (Macmillan, New York, 2016).

Beard, Jo Ann, *The Boys of My Youth* (Phoenix House, London, 1998).

Bekoff, Marc and Pierce, Jessica, *Wild Justice: The Moral Lives of Animals* (University of Chicago Press, Chicago, 2009).

Berger, John, *Why Look at Animals* (Penguin, London, 2009).

——, *Parting Shots from Animals*, a BBC Omnibus (1980) documentary, based on essays by John Berger, directed by Michael Dibb and Christopher Rawlence.

Boehrer, Bruce Thomas, *Parrot Culture: Our 2500-year-old Fascination with the World's Most Talkative Bird* (University of Pennsylvania Press, Philadelphia, 2004).

Braidotti, Rosa, *The Posthuman* (Polity Press, Cambridge, 2013).

Braithwaite, Victoria, *Do Fish Feel Pain?* (Oxford University Press, Oxford, 2010).

Caffo, Leonardo, *Only for Them: A Manifesto for Animality Through Philosophy and Politics* (Mimesis International, 2014).

Calarco, Matthew, *Zoographies: The Question of the Animal from Heidegger to Derrida* (Columbia University Press, New York, 2008).

Carson, Rachel, *Silent Spring* (Penguin Classics, London, 2002).

Cartmill, Matt, *A View to a Death in the Morning: Hunting and Nature Through History* (Harvard University Press, Cambridge, MA, 1993).

Chignell, Andrew, Terence Cuneo, and Matthew C. Halteman (eds.), *Philosophy Comes to Dinner: Arguments on the Ethics of Eating* (Routledge, New York, 2016).

Clutton-Brock, Juliet, *A Natural History of Domesticated Mammals*, 2nd edition (Cambridge University Press, Cambridge, 1999).

Coates, Peter, *Strangers on the Land: American Perceptions of Immigrant and Invasive Species* (University of California Press, Berkeley, 2007).

Cronon, William, *Nature's Metropolis: Chicago and the Great West* (W.W. Norton, New York, 1992).

D'Arcy Wentworth Thompson, *On Growth and Form* (Cambridge University Press, Cambridge, 1997).

Davis, Mark A., *Invasion Biology* (Oxford University Press, 2009).

Deleuze, Gilles and Guattari, Félix, 'Becoming Animal', in Linda Kalof and Amy Fitzgerald (eds.), *The Animals Reader: The Essential Classical and Contemporary Writings* (Berg, Oxford, 2007).

Derrida, Jacques, *The Animal That Therefore I Am*, Marie-Louise Mallet (ed.), trans. David Wills (Fordham University Press, New York, 2008).

Descartes, René, 'Discourse on the Method of Conducting One's Reason Well and of Seeking the Truth in the Sciences', in George Heffernan (ed.), trans. George Heffernan, *Discourse on the Method* (University of Notre Dame Press, Notre Dame, 1637/1994).

Despret, Vinciane, *What Would Animals Say if We Asked the Right Questions?* (University of Minnesota Press, Minneapolis, 2016).

de Waal, Frans, *Are We Smart Enough to Know How Smart Animals Are?* (Granta, London, 2017).

——, *Mama's Last Hug* (Granta, London, 2019).

Dombrowski, Daniel, *The Philosophy of Vegetarianism* (University of Massachusetts Press, Mass., 1984).

Douglas, Mary, *Purity and Danger* (Routledge, London and New York, 2002).

Eiseley, Loren, *The Immense Journey* (Vintage, New York, 1957).

——, *The Night Country* (Charles Scribner's Sons, New York, 1972).

——, *Notes of An Alchemist* (Charles Scribner's Sons, 1972).

——, *All the Strange Hours: The Excavation of a Life* (Charles Scribner's Sons, New York, 1975).

Emberly, Julia V., *The Cultural Politics of Fur* (Cornell University Press, Ithaca, 1998).

Fine, Cordelia, *Testosterone Rex: Unmaking the Myths of Our Gendered Minds* (Icon, London, 2018).

Francione, Gary L., *Rain without Thunder: The Ideology of the Animal Rights Movement* (Temple University Press, Philadelphia, 1996).

Fudge, Erica, *Animal* (Reaktion, London, 2002).

——, *Pets: Art of Living* (Acumen, Stocksfield, 2008).

Gould, Stephen Jay, *Wonderful Life: Burgess Shale and the Nature of History* (Vintage, London, 2000).

——, *I Have Landed: Splashes and Reflections in Natural History* (Vintage, London, 2003).

Grandin, Temple and Johnson, Catherine, *Animals Make Us Human: Creating the Best Life for Animals* (Mariner Books, Boston, 2010).

Griffin, Donald R., *Animal Thinking* (Harvard University Press, Cambridge, Mass., 1984).

Gruen, Lori (ed.), *The Ethics of Captivity* (Oxford University Press, New York, 2014).

Guerrini, Anita, *Experimenting with Humans and Animals: From Galen to Animal Rights*, Johns Hopkins Introductory Studies in the History of Science (The Johns Hopkins University Press, Baltimore, MD, 2003).

Haraway, Donna, *The Companion Species Manifesto: Dogs, People, and Significant Otherness* (Prickly Paradigm Press, Chicago, IL, 2003).

——, *When Species Meet* (University of Minnesota Press, Minneapolis, 2007).

Horkheimer, Max, *The Eclipse of Reason* (Oxford University Press, Oxford, 1947).

Hubert, M. and Mauss, M., *Sacrifice: Its Nature and Functions* (Midway Reprint, Chicago, 1981).

Ingold, Tim, 'Humanity and Animality', in Tim Ingold (ed.), *Companion Encyclopedia of Anthropology*, 14–32 (Routledge, New York, 1994).

—— (ed.), *What is an Animal?* (Routledge, London, 1994).

——, *Lines, A Brief History* (Routledge, London, 2007).

Kalof, Linda (ed.), *A Cultural History of Animals in Antiquity*, Vol.1 (Berg, London and New York, 2011).

Kellert, Stephen R. and Wilson, Edward O. (eds.), *The Biophilia Hypothesis* (Island Press, Washington, DC,1993).

Kheel, Marti, *Nature Ethics: An Ecofeminist Perspective* (Rowman & Littlefield, Lanham, 2008).

King, Barbara J., *How Animals Grieve* (University of Chicago Press, Chicago, 2013).

Latour, Bruno, *We Have Never Been Modern*, trans. Catherine Porter (Harvard University Press, Cambridge Mass., 1993).

Leopold, Aldo, *A Sand County Almanac* (Oxford University Press, Oxford, 1949).

Linzey, Andrew, *Why Animal Suffering Matters* (Oxford University Press, Oxford, 2009).

Linzey, Andrew and Clair Linzey (eds.), *The Palgrave Handbook of Practical Animal Ethics* (Palgrave Macmillan, London, 2018).

Lopez, Barry, *Crossing Open Ground* (Vintage, New York, 1989).

Lorenz, Konrad, *Man Meets Dog* (Methuen, London,1954).

——, *On Aggression* (Routledge, London, 2002).

Lorca, Federico García, *Poet in New York*, (Penguin, London, 2002).

Malamud, Randy, *Reading Zoos: Representations of Animals and Captivity* (New York University Press, New York, 1998).

Manguso, Sarah, *The Guardians: An Elegy* (Granta, London, 2002).

Martin, Martin, *A Late Voyage to St Kilda*, first published in 1698, reprinted by Franklin Classics, 2018.

Midgley, Mary, *Animals and Why They Matter* (University of Georgia Press, Athens, GA, 1993).

——, *Beast and Man: The Roots of Human Nature* (Routledge, New York, 1995).

Mills, Brett, *Animals on Television: The Cultural Making of the Non-Human* (Springer, New York, 2017).

Modiano, Patrick, *The Search Warrant* (Harvill, London, 2000).

Nelson, Richard K., *Make Prayers to the Raven: A Koyukon View of the Northern Forest* (University of Chicago Press, Chicago,1983).

Noske, Barbara, *Humans and Other Animals: Beyond the Boundaries of Anthropology* (Pluto Press, London, 1989).

Oliver, Mary, *Upstream: Selected Essays* (Penguin Press, New York, 2016).

Osborne, Catherine, *Dumb Beasts and Dead Philosophers: Humanity and the Humane in Ancient Philosophy and Literature* (Oxford University Press, Oxford, 2007).

Packham, Chris, *Fingers in the Sparkle Jar: A Memoir* (Ebury Press, London, 2017).

Paton, William, *Man and Mouse: Animals in Medical Research* (Oxford University Press, Oxford, 1993).

Pick, Anat, *Creaturely Poetics: Animality and Vulnerability in Literature and Film* (Columbia University Press, New York, 2011).

Pierce, Jessica, *Run, Spot, Run: The Ethics of Keeping Pets* (University of Chicago Press, Chicago, 2016).

Poliquin, Rachel, *The Breathless Zoo: Taxidermy and the Cultures of Longing* (Pennsylvania State University Press, Pennsylvania, 2012).

Regan, Tom, *All That Dwell Therein: Essays on Animal Rights and Environmental Ethics* (University of California Press, Berkeley, 1982).

——, *The Case for Animal Rights* (University of California Press, Berkeley, 1983).

——, *The Struggle for Animal Rights* (International Society for Animal Rights, Clarks Summit, PA, 1987).

——, *Defending Animal Rights* (University of Illinois Press. Urbana, IL, 2001).

——, *Animal Rights, Human Wrongs: An Introduction to Moral Philosophy* (Rowman & Littlefield, Lanham, MD, 2003).

——, *Empty Cages: Facing the Challenge of Animal Rights* (Rowman & Littlefield, Lanham, MD, 2004).

Regan, Tom and Singer, Peter (eds.), *Animal Rights and Human Obligations* (Prentice-Hall, Englewood Cliffs, 1976).

Rothschild, Miriam, *Animals and Man* (Clarendon Press, Oxford, 1986).

Safina, Carl, *Beyond Words: What Animals Think and Feel* (Henry Holt, New York, 2015).

Sagan, Carl, Druyan, Ann, *Shadows of Forgotten Ancestors: A Search for Who We Are* (Century, New York, 1992).

Shrubsole, Guy, *Who Owns England?: How We Lost Our Green and Pleasant Land and How to Take it Back* (Collins, London, 2019).

Sinclair, Upton, *The Jungle* (Penguin, London, 2002).

Singer, Peter, *Animal Liberation* (Pimlico, London, 1995).

Solnit, Rebecca, *The Mother of All Questions* (Granta, London, 2017).

Sorabji, Richard, *Animal Minds and Human Morals: The Origins of the Western Debate* (Cornell University Press, Ithaca, 1993).

Thomas, Keith, *Man and the Natural World: Changing Attitudes in England 1500–1800* (Penguin Books, London, 1983).

Tüür, Kadri and Tønnessen, Morten (eds.), *The Semiotics of Animal Representations* (Rodopi, Amsterdam, 2014).

Winton, Tim, *Island Home* (Picador, London, 2017).

Wright, Thomas, *Circulation: William Harvey's Revolutionary Idea* (Vintage, London, 2013).

Index